Deliberating Environment Policy in India

As one of the world's largest and most bio-diverse countries, India's approach to environmental policy will be very significant in tackling global environmental challenges. This book explores the transformations that have taken place in the making of environmental policy in India since the economic liberalization of the 1990s. It investigates if there has been a slow shift from top-down planning to increasingly bottom-up and participatory policy processes, examining the successes and failures of recent environmental policies. Linking deliberation to collective action, this book contends that it is crucial to involve local actors in framing the policies that decide on their rights and control over bio-resources in order to achieve the goal of sustainable human development.

The first examples of large-scale participatory processes in Indian environmental policy were the 1999 National Biodiversity Strategy Action Plan and the 2006 Scheduled Tribes and Other Traditional Forest Dwellers Act. This book explores these landmark policies, examining the strategies of advocacy and deliberation that led to both the successes and the failures of recent initiatives. It concludes that in order to deliberate with the state, civil society actors must engage in forms of strategic advocacy with the power to push agendas that challenge mainstream development discourses. The lessons learnt from the Indian experience will not only have immediate significance for the future of policy making in India, but they will also be of interest for other countries faced with the challenges of integrating livelihood and sustainability concerns into the governance process.

Sunayana Ganguly is a Research Associate, Industrial Ecology Group, University of Lausanne, Switzerland and Post-Doctoral Fellow, Ashoka Trust for Research in Ecology and the Environment (ATREE), Bangalore, India.

Routledge Studies in Asia and the Environment

The role of Asia will be crucial in tackling the world's environmental problems. The primary aim of this series is to publish original, high-quality, research-level work by scholars in both the East and the West on all aspects of Asia and the environment. The series aims to cover all aspects of environmental issues – including how these relate to economic development, sustainability, technology, society, and government policies – and aims to include all regions of Asia.

Deliberating Environment Policy in India

Participation and the role of advocacy

Sunayana Ganguly

Routledge
Taylor & Francis Group

LONDON AND NEW YORK

First published 2016 by Routledge

2 Park Square, Milton Park, Abingdon, Oxfordshire OX14 4RN

711 Third Avenue, New York, NY 10017

Routledge is an imprint of the Taylor & Francis Group, an informa business

First issued in paperback 2017

British Library Cataloguing-in-Publication Data
A catalogue record for this book is available from the British Library

Library of Congress Cataloging-in-Publication Data
Names: Ganguly, Sunayana, author.
Title: Deliberating environment policy in India : participation and the role of advocacy / Sunayana Ganguly.
Description: New York, NY : Routledge, 2016. | Series: Routledge studies in Asia and the environment | Includes bibliographical references.
Identifiers: LCCN 2015021311
Subjects: LCSH: Environmental policy—India. | Environmental policy—India—Citizen participation. | Sustainable development—India. | India—Economic policy—1991– | India—Economic policy—Environmental aspects.
Classification: LCC HC440.E5 G37 2016 | DDC 333.70954—dc23LC record available at http://lccn.loc.gov/2015021311

ISBN: 978-1-138-81943-6 (hbk)
ISBN: 978-1-138-47673-8 (pbk)

Typeset in Times New Roman
by Apex CoVantage, LLC

Contents

4   Deliberating on the Forest Rights Act   94

5   Conclusion   132

Illustrations

Figure

Tables

Abbreviations

AAP	Aam Aadmi Party
BJP	Bharatiya Janata Party
BSAP	Biodiversity Strategy Action Plan
CBD	Convention on Biological Diversity
CFP	call for participation
CMP	Common Minimum Programme
CPI(M)	Communist Party of India (Marxist)
CSD	Campaign for Survival and Dignity
CSO	Civil Society Organization
CWH	Critical Wildlife Habitat
FRA	Scheduled Tribes (Recognition of Forest Rights) Act
GEF	Global Environment Facility
GoM	Group of Ministers
IPR	intellectual property rights
JFM	joint forest management
JPC	Joint Parliamentary Committee
MoEF	Ministry of Environment and Forests
MoTA	Ministry of Tribal Affairs
NBSAP	National Biodiversity Strategy and Action Plan
NBWL	National Board for Wildlife
NDA	National Democratic Alliance
NFFPFW	National Forum of Forest People and Forest Workers
NGO	non-governmental organization
NTFP	non-timber forest produce
PA	protected areas
TPCG	technical and policy core group
TRIPS	trade-related aspects of intellectual property rights
TSG	Technical Resource Group
UNDP	United Nations Development Programme
VO	voluntary organizations
VSS	*Van Suraksha Samitis*
WLPA	Wildlife Protection Act
WTO	World Trade Organization

Acknowledgements

I would like to start with expressing my most sincere gratitude to everybody who has been part of this journey. This book has emerged from the doctoral research I carried out at the Freie Universität Berlin, and it has been a true experience of bridging East and West, with all its contradictions, frustrations and joy. While it is impossible to detail the contributions of every individual, and many have remained unacknowledged by name, I would like to thank this large ecosystem of people spanning several countries for their unflinching support and generosity. I hope this study does justice to their contributions.

First and foremost I would like to thank Dr. Miranda Schreurs. This book would not have been possible without her guidance, support, editing and advice. I am also grateful to Dr. Frank Fischer, whose comments allowed me to fine-tune this manuscript, and for the encouragement of my dissertation committee, who urged me to publish. I would especially like to mention Dr. Gunnel Cederlöf and Dr. Beppe Karlsson, who kindly found the time to give me inputs on my work at a crucial stage in the research process.

This research began alongside my work at the Deutsche Institut für Entwicklungspolitik (DIE), Bonn, and I could not have completed it without their support – both financial and otherwise – especially the department of "Competitiveness and Social Development," in particular Andreas Stamm, Tilman Altenburg, Dirk Messner and colleagues and friends without whom this dissertation would never have been written.

I am very grateful to my interviewees in India whose experience, insights and passion made fieldwork such an adventure. Without the insights of (in alphabetical order) Dr. Parthasarathi Banerjee, Dr. Ravi Chellam, Ashok Chowdhury, Madhav Gadgil, Kanchi Kohli, Ashish Kothari, Professor Kailash Malhotra, Dr. Nagaraja, Sudarshan Rodriguez, Madhu Sarin, Shekhar Singh, Darshan Shankar, Tejaswini Apte and Sanjay Upadhaya, the texture of this research would have been very different. Their important and inspiring work in these areas combined with their willingness to help and discuss different points of view was invaluable.

Thank you to Baptiste Fabre for his support through this process. A big thank you to all of my friends – from India, Germany and France – without whom this process would have been a far lonelier undertaking. With discussions, encouragement, advice and commiserative drinking, they shared with me the daily trials and triumphs of the writing process and kept me alive to the possibilities of the world beyond.

I would also like to extend my gratitude to my parents (Kalyan and Suparna Ganguly), my sister (Suranjana) and my nani (Sushanta Baksi), and to Uma Anand, Radnyee Pradhan and Amena Zavery and the extended family clans who still do not know exactly what my research consists of but whose love, support and humor have been the foundations of my emotional life.

Finally, I am very thankful to Routledge and my editors Helena Hurd and Peter Sowden for their valuable insights, comments and encouragement.

1 Introduction

Those on the margins often come to control the center. And those in the center make room for them, willingly or otherwise.

Daxos, *Game of Thrones*

Environment policy is a space that encompasses different constructions of "nature," "economy" and "livelihood" that contain tensions among different sets of social and ethical concerns, economic or political preferences, scientific and technological systems, traditions and knowledge. Given these cross-cutting characteristics, environmental policy is important as a space where competing narratives of economic and social processes can be revealed. In particular, the governance of environmental resources in India negotiates between different meanings, identities and political imaginaries that are thrust on it. This negotiation is the core element of environmental governance, which is the way societies deal with environmental problems. It is concerned with interactions and interdependencies among formal and informal institutions and actors within society that influence how environmental governance processes unfold, problems are defined, issues are framed and contention is dealt with.

The space for these negotiations is captured in Partha Chatterjee's notion of "political society" (Chatterjee 2001; Kaviraj and Khilnani 2001) which points out that in practice the politics of democracy is frequently carried out not in civil society associations but in legally ambiguous, mediating spaces between the state and civil society, especially in post-colonial democracies (Leach et al. 2007, 10–11). An extensive body of work (Mathur and Mathur 2007; Holston and Appadurai 1999; Sharma 1999) concurs that social forces profoundly influence the state in India and that the boundaries between the state and society are porous and blurred with ordinary people recognizing that this is so. However, comparatively less research has attempted to capture these ideas with respect to policy-making, which continues to be conceptualized as an activity confined to the jurisdiction of the state. If sustainable human development is the ultimate goal, then resilient ecosystems and communities are essential in bridging disparities in economy, equity and opportunity. Biodiversity and forests, considered essential public goods and natural provider of ecosystem services, remain undervalued in national ecosystem services and are thus given much less space in policies. Loss of biodiversity and forest cover has severe consequences for poverty reduction and livelihood security, as it is the poor and vulnerable who are directly dependent on these resources for their subsistence needs, both as income and as insurance.

National development choices affect the control and quality of biodiversity and forests and define the winners and losers of the development process. In the nexus of climate change, industrialization, habitat loss and overexploitation of resources, the most affected communities are those that are directly dependent on natural or semi-natural ecosystems. Yet the way development processes are organized and governed often neglect this reality. The role of governance should be to carve out a role for these communities, who should participate in framing policies that decide on their rights and control over bio-resources. In this, it is critical to examine the state's role and interactions with society in the process of policy-making. The state can no longer be viewed as a monolithic actor in the operationalization of its practices and institutions. Instead within the policy-making process, the boundaries between state and society are re-interpreted and re-constituted through the emergence of participatory spaces and the formation of networks and movements that cross state-society divides.

The idea of "participation" has gained political currency in bridging the state-society divide with inclusive governance becoming the preferred prescription. In the academic world, participation and deliberation have been treated from a variety of different lenses. The failures of traditional representative democracy have been studied as a decline in levels of associational life in mature democracies (Putnam 2000; Skocpol and Fiorina 1999) and the growing influence of narrow interest groups (Olson 1965; Cohen and Arato1992). In the developing world, the idea of participation has come to be seen as critical in terms of increasing the capabilities and rights of citizens (Drèze and Sen 1995) as well as strengthening the "voice" of the people traditionally marginalized from policy processes. This emphasis on inclusion in participatory models of democracy overlaps with deliberative and discursive democrats for whom social inclusion is inherent to a successful democracy. However, there is particularly a major gap in the literature on what strategies civil society organizations use to carve out spaces for different strategies of political engagement and inclusion. India has a rich tradition of social mobilization that has left an imprint on the politics of the country – whether they are environmental or social movements, anti-corruption movements or anti-rape movements. But few attained the scale and influence to impact the political arena. Some exceptions to this are the farmers' movements of the 1980s and the movement around the "right to information" in the 1990s, but, in general, social movements have created change from beyond the purview of the state.

The growing idea of participation of the common person in the everyday decisions of governance is a relatively new phenomenon. Although India has traditionally subscribed to a system where government officials or insiders make policy decisions, the last decade has seen a growing demand for inclusive forms of governance. One sees a growing dynamism in policy-making in India, which does not rely on authoritative decision-making but is rooted in more participatory, deliberative processes in which boundaries between the notions of state and society are more "elusive, porous and mobile" (Mitchell 1991, 77). The idea of the Indian state as a monolithic structure has been extensively critiqued (Rai 1996; Sharma 1999), and emerging points of view underline the increasing incorporation of a plurality of voices, actors and institutions (Chopra 2011). State-level processes have opened structures to a greater variety of voices as civil society has begun pushing for the opening of more deliberative spaces within the state.

It has been a popular view that environmental governance in India emerged as a response to global initiatives and was a result of external initiatives rather than any long-term vision of domestic policymakers (Dwivedi 1997; Pal 1999). Other scholars like Sinha (1998) argue that demands for governance have evolved in the context of social movements in India. They emerged as a response to the resource-oriented development pattern that diminished productive potentiality of natural resources and created severe ecological imbalances. Similarly, scholars like Kohli (1994) argue that the evolution of governance of natural resources in India is because of the emergence of societal forces in general and environmental movements in particular. A large part of the literature on environmental governance in India has focused on social movements (Shiva 1991; Kothari 2001) with little work existing on the role of the state in negotiating and balancing competing mass interests within a policy process (Chopra 2011). An emerging literature is focused on the changing role of the state because of the "porous interactions between the state and society during the policy-making process" (Chopra 2011, 91). However, this literature does not explicitly address the role and strategies of advocacy with relation to the changing nature of the state. There is a huge gap in the rhetoric and reality of how collective action is shaped in India (Deo and McDuie-Ra 2011), as well as the role of the state in negotiating with different demands placed on it and the blurred spaces where the negotiation takes place.

According to Anil Agrawal (1994, 346), the environment is "an idea whose time has come in India." It has been argued that policy analysis in India, because of its post-colonial predilection on "planning," has been undertaken through the lenses of economic planning and public administration. More recently the growth of independent research institutions that play the twin roles of analysis and advocacy has led to a reframing of public issues. These institutions exploit the space between "the state and the sphere of civil society organizations rooted in a participatory politics" (Mathur and Mathur 2007, 603). This re-conception in the analysis of the political changes in India, moving from a dominant economic policy analysis to the capturing of the more dynamic political struggles for more democratic functioning, has been particularly important in the studies on environment struggles (Guha 1989; Baviskar 1995; Drèze et al. 1997). These studies mainly focus on social movements outside the state in India (Shah 2004; Dwivedi 1997), while little is written on movements coopted into the state. In particular, very little literature exists on how organizations and people are promoting participation within deliberative spaces and therefore transforming the potential for participation by formulating policy with and by the marginalized sections of the country. It is to this strategic engagement with the state and dynamic interactions that policy research has begun to pay more theoretical attention.

There appears to be a growing interest in alternative discourses, the role of civil society and social movements (Shah 1991), struggles over resources (Baviskar 1997, among others) and advocacy in the Indian context (Deo and McDuie-Ra 2011). There are good reasons for testing the theoretical models of deliberation in countries that institutionally discourage deliberation and adapting these models to different patterns of interaction. A majority of the world's population lives in poor, democratizing, divided societies, where culturally and socially appropriate demo-cratic institutions are desperately needed, especially in light of the marginalization that these populations face in establishing their voice and interests in the policy process. Yet most contemporary political and social scientists use developed and industrialized countries to test the theory of deliberative democracy (Chambers 2003, 318), where the conditions for instituting the ideals of deliberative experi-ments are relatively favorable. If policy-making can be recognized as something emerging from a networked society rather than a hierarchical system (Hajer and Wagenaar 2003), then one must focus not only on the state that has the ultimate decision-making authority but also on the processes of strategic engagement with the state undertaken by civil society actors. In recent years, there has been a move to study the deliberative turn in politics in India (Fischer 2006; Heller 2001; Ban et al. 2012). The complex relationships between a multitude of actors – including state institutions, civil society and international organizations – are important in order to understand the patterns of conflicting agendas and discourses in order to have a fuller understanding of the Indian policy process.

The questions that guide understandings of deliberation and civil society advo-cacy are best captured in the process of policy formulation. Capturing the diversity of voices, underlining the co-optation of marginalized discourses and conferring legitimacy on policies in the context of environment policy formulation in India is best discussed in the first stage of policy development which has largely remained encapsulated in a "black box" (Corkery et al. 1995, 2). The fundamental prob-lem in the conceptual framework of the policy process is that it is seen as a lin-ear model (Linder and Peters 1989). It does not interrogate the complexities of policy-making in its entirety but rather is reduced to a sequence of steps, with an identifiable beginning and an end. This framework assumes that policy formula-tion encompasses rational tradeoffs, incentives, bargaining and optimal choices to suit existing conditions. It fails to capture the subjective and dynamic interactions between state and non-state actors, the values and beliefs of the actors involved, the strategic advocacy and ensuing conflicts in narratives and discourses. This analysis moves away from an objective, positivist, value neutral understanding of policy analysis to focus on the messy interactions and dynamic by which policy is created and negotiated.

The purpose of this book is twofold: the first is to explore the role of policy advocacy on behalf of, and by, marginalized sections of society and the role of civil society organizations who push rights-based claims; the second is to con-sider their strategies in pushing for a more deliberative and participative model of policy formulation. Comparing the success of one process, the formulation of the Forest Rights Act (FRA), with the failure of another, the formulation of the National Biodiversity Strategy and Action Plan (NBSAP), highlights this. The comparison shows that in order to negotiate within an institutional or state-led deliberative space and arrive at a more participatory policy process, civil society actors must engage in extra-parliamentary forms of strategic advocacy. This was seen in the case of the Forest Rights Act where organized mobilization of the grassroots ultimately found representation within the deliberative sphere of the state. An incremental build-up of structures of mobilization allowed organizations to tailor their responses to political opportunities provided by the state. They could enter forums of deliberation when required and used extra-parliamentary tactics of mobilization and public pressure when the government threatened to retreat. In contrast, the NBSAP was spearheaded by a government that initiated a deliberative process. Advocacy entered the process quite late, after the government had already retreated. Structures that were mainly oriented towards deliberation could not then transform into processes of advocacy. Thus it ultimately led to a deliberative failure as well.

Some of the main actors in this negotiation are voluntary organizations, non-governmental organizations, experts, activists, scientists and others, whom I collectively refer to as civil society actors and organizations. These organizations advocate for more inclusive spaces for policy deliberation that encourage public debate and consensual decision-making. I compare two processes of interaction with the state where civil society actors exploit policy windows to push for more participative forms of deliberation in policy formulation. The first case is centered on biodiversity. It portrays "invited spaces": spearheaded by government institutions and actors who introduce new institutions and models of public participation. This case illustrates the role that the Convention on Biological Diversity played in advancing particular ideas and modes of governance (e.g. participatory processes). These ideas were then advocated for in the national arena by civil society organizations and legitimized by donors and national governments. I refer to these processes as "top-down" processes given that they are initiated by the state and then trickle into civil society. The second case focusing on forest rights portrays "invented spaces" that explain modes of protest and mobilization by civil society organizations who occupy available spaces or create new spaces for participation. They advocate for more effective deliberation within the policy spaces or forums provided by the state and link it to mass mobilization in the public sphere. These are referred to as "bottom-up" processes given that they are initiated by civil society and then find representation in the state processes of policy formulation.

The two cases were chosen not for their representativeness but rather to contrast the advocacy strategies found within two different openings for deliberation within the Indian policy context. This is based on the idea that the behavior of civil society organizations is influenced not only by the institutional set-up that they are embedded in, but also the policy context and cultural norms that surround them. This approach allows comparisons to be made between different coalitions in the same policy space. The cases were selected on the basis of their diversity, age, location and scope, and deliberative features. The cases cover a diversity of issues, organizations and processes. They both deal with issues that are contested and complex and involve a multitude of actors who are all following their own financial, social, ethical and scientific ideas. Finally, both cases had the features of mass participation, advocacy at different levels and inclusion of marginalized voices. In addition, they were involved in both formal and informal spaces of deliberation sponsored by a variety of policy actors, including government, research organizations, social movements and international organizations, which are further explained in the case study. The two cases, the NBSAP and FRA, allow me to analyze the changes and shifts in dynamics, strategies and actors in the formal and informal policy space as well as the historical and contemporary events that shaped their advocacy strategies.

There are several pathways to more participatory forms of governance. These pathways are embodied in the two policy processes where the strategies of civil society advocacy include the public framing of issues, deriving legitimacy through international norms, the formation of effective civil society alliances through the mobilization of grassroots actors and the creation of disaggregated forums to capture different worldviews of the resource being discussed. The outcomes of two case studies focused on rights related to biodiversity and forestland. This provides an important framework by which to understand the changing nature of policy processes in India, particularly people's participation and demands for certain rights and the responses to these rights-based demands within the traditionally closed policy arenas that have long dominated the Indian policy landscape.

Environmental planning processes in India

Environmental institutions have historically limited large-scale participation, and the emergence of the environmental agenda in India evolved incrementally post-independence in 1947. Post-independent India followed the Soviet-style model of long-term centralized planning and adopted a system of five-year plans. In 1952, the Planning Commission declared, "[T]he central objective of planning in India is to raise the standard of living of the people and to open out to them opportunities for a richer and more varied life" (Khator 1991,11). This has led to a growth paradigm that is embraced by government leaders even today. As P. Chidambaram (2006), the Indian finance minister, stated unequivocally, the government of India is "willing to tolerate debate, and perhaps even dissent, as long as it doesn't come in the way of 8 percent growth" (*Hindu* 2006c). The environmental challenge was initially seen as a threat to economic development in India. This is still a prevalent view in some quarters of the government, who consider it merely another northern country concern, constructed to keep the world poor. J. Mohan Rao (2005, 681) explains, "[M]any in India today including government officials . . . regard the environmental lobby as a child of northern conspiracy and northern funding."[1]

The objective of central planning remained unchanged for decades; nevertheless there were shifts from industrialized growth to agricultural growth reflecting the changing demands and objectives of the nation. Once the nation's economy slowly stabilized, the issues of fair and equitable distribution of resources made an appearance on the agenda. This was particularly emphasized by the success of the Green Revolution[2] in 1965. Consequently, with growth the government was pressured into devoting resources to the social agenda. The national agenda however did not make any space for the environmental agenda, even though serious environmental problems were already surfacing by the 1950s. It was in the Fourth Five-Year Plan (Planning Commission 1966) that the issue of environmental protection was for the first time stated explicitly within an official policy document. Within its long-term perspective, it stated:

> Planning for harmonious development recognizes this unity of nature and man. Such planning is possible only on the basis of a comprehensive appraisal of environmental issues; particularly economic and ecological. . . . [I]t is necessary, therefore, to introduce the environmental aspect into our planning and development.
>
> (Planning Commission 1969, chap. 2)

It was non-committal but sufficient to pave the way for more serious engagement (Khator 1991, 11).

Serious engagement with environmental issues in India remained negligible until 1971, when the Planning Commission wrote reports on the state of India's environment in preparation for the 1972 United Nations Conference on the Human Environment. By May 1971, three reports had been prepared: "Some Aspects of Environmental Degradation and Its Control in India," "Some Aspects of Problems of Human Settlement in India" and "Some Aspects of Rational Management of Natural Resources." By 1972, it was decided that it was necessary to establish a central coordinating body to integrate environmental concerns with larger plans of economic development; thus the National Committee on Environmental Planning and Coordination (NCEPC) was created in the Department of Science and Technology. This was an apex advisory body whose mandate was to co-ordinate all matters relating to environmental protection and improvement. However with over-bureaucratization leading to ineffective decision-making, the power of the NCEPC eventually waned. The Fifth Five-Year Plan (Planning Commission 1974, foreword–chap. 2) continued to emphasize the link between poverty and environmental degradation. It also stressed the importance of taking environmental goals into account in all industrial decision-making and maintaining the balance between development planning and environmental management (Divan and Rosencranz 2001, 34).

Various changes also took place in the institutional set-up. In 1985 as the NCEPC slowly weakened, the government created the Ministry of Environment and Forests so that a unified administrative set-up could coordinate the various aspects related to natural resource management. This was highlighted in the Sixth Five-Year Plan that pointed out:

> Plans and programmes in fields of soil conservation, public health, forest and wildlife protection, industrial hygiene etc. have been in existence in India for many decades. However the first formal recognition of the need for integrated environmental planning was made when the Government of India consti-tuted the National Committee on Environmental Planning and Coordination (NCEPC) in 1972.
>
> (Planning Commission 1980, 20.38)

At the end of 1982, the status of this ministry was upgraded, with the appoint-ment of a senior minister with the rank of cabinet minister. The passage of the Environmental Protection Act in 1986 sealed the institutional profile of the min-istry (Dwivedi and Khator 1995, 51). In addition to other activities, it is also in charge of setting up linkages with international programs and participating and formulating international conventions and treaties like the Convention on Biological Diversity and the Montreal Protocol. In keeping with this profile, the scale and budget of the ministry has also expanded over the years.

India has historically institutionalized its environmental concerns, though this is linked to colonial demands for timber and natural resources. Prior to coloniza-tion, most natural resources were still loosely under the centralized control of monarchs. They largely left the local people's customary rights undisturbed in their access and utilization of resources for daily subsistence. As discussed by Gadgil and Guha (1996), for many centuries a culture of decentralized partici-patory decision-making (through village and community units and leaders) and a careful system of natural resource management formed the basis of common resource use and conservation strategies. With the advent of colonization, there was a weakening of these social relationships and community management sys-tems. With the British came a commodification of resources as well as a more centralized and bureaucratic model of state control of natural resources, which were until then unknown in the subcontinent.

The first legal attempt at asserting state control of resources in colonial India was through the Indian Forest Act of 1865. In 1889, Baden-Powell, a senior civil servant in the British government, claimed that "the right of the State to dispose of or retain for public use the waste or forest area is among the most ancient and undisputed features of oriental Sovereignty" (quoted in Dubey 2006, para. 3) – justifying British control over these resources and alienating the local community from the decision-making processes (Dubey 2006). Following the colonial tradi-tion over decades, India has retained a very centralized approach to environmental governance. It prescribed limited roles for states and sub-states with the Ministry of Environment and Forests retaining complete control over all areas pertaining to the environment.

Until the Fourth Five-Year Plan, the environmental agenda was definitely undermined by developmental agenda, although environmental issues slowly began to gain momentum in the consequent plans. In the Sixth (1980–85), Seventh (1985–90) and Eighth (1992–97) Five-Year Plans, the tone fluctuated between the alarmist call of the early eighties to the gradual recognition of participation in environmental management. In the Sixth Five-Year Plan, the chapter on environment (Planning Commission 1980, chap. 20) highlighted degradation, damage and loss of species as growing concerns. It also for the first time mentioned the "Conservation of Genetic Resources and Natural Ecosystems," which stated:

> The diversity of biological organisms is a vital resource which needs to be carefully protected in natural ecosystems [if we are not to close many possible evolutionary options for benefiting future generations]. These natural ecosystems [are] a vitally important economic resource.
>
> (20.17)

The Seventh Five-Year Plan highlighted the negative effect development programs were having on the environment. It reflected on the fact that "the need to improve the conditions of our people is pressing; under this pressure many concerned with developmental activities lose sight of environmental and ecological imperatives" (Planning Commission 1985, 18.2). For the first time, the plan called out for people working together to improve the quality of the environment and highlighted "securing greater public participation in environmental management" (Planning Commission 1985, 18.20). Before this there was no formal bid to include people in the management of natural resources; the idea had been to focus on people-centered development and that the environment would follow. Though implementation of this principle was negligible, it is the first recognition of the fact that effective participation is required in environmental management and conservation. The Eighth Five-Year Plan (Planning Commission 1992, chap. 4) attempted to integrate environmental concerns sectorally and aimed to reconcile the conceptual conflict between development and environment. It also put biodiversity on the table but listed it as a global concern, keeping it distanced from the national agenda. In preparation for the Rio conference of 1992, the government stated:

> Environmental issues such as depletion of Ozone layer, Greenhouse gases and climate change, bio-diversity and role of forests are current global concerns. Some of these issues are to be discussed shortly at United Nations Conference on Environment and Development to be held in Brazil in June 1992. It is essential that these negotiations recognise the aspirations of large masses of poor people and do not impose any burden on developing countries, respecting their sovereign right over their resources. Transfer of technology, flow of new and additional resources to developing countries to fully meet any additional cost, are pre-requisites to international cooperation in the environment sector.
>
> (Planning Commission 1992, 4.16.7)

Article 253 of the India Constitution, for instance, empowered parliament to make any law for the whole country to implement the decision taken at international conferences even for the subjects under the jurisdiction of the states (Constitution of India, art. 253, entries 13–14, list I, schedule VII). Because India was a signatory to the United Nations Stockholm Declaration in 1972, the central government claimed sole jurisdiction over environmental matters based on the fact that environmental rules were derived from international obligations. The 1976, the forty-second amendment to the India Constitution moved the subjects of "forests" and "protection of wild animals and birds" from the State List to the Concurrent List, which meant that it could be legislated by both the center and the state, which meant that in any conflict of interest, the center would prevail. It was under these frameworks that the Air Act of 1981 and the Environment Protection Act of 1986 were enacted without public debate and minimum involvement of the states (Dubey 2006, para. 8).

During the 1980s, the direction of environmental policy took "Indian federalism simultaneously toward legal centralism and administrative devolution" (Sivaramakrishnan 2003). This was a result of the complex legacy of colonialism as well as many national developments in the politics of the 1980s. Indian federalism (developed more as a convenient administrative unit rather than the pledge of independent units) allowed for powers to be granted to states by the center. In addition, the various economic and foreign crises the nation faced after independence made it necessary to keep the center strong. During the rule of Indira Gandhi (1966–77, 1980–84), the drive towards centralization reached its peak. However, instead of settling into a unitary system, it led to a lopsided growth where the states' abilities decreased while their responsibilities kept increasing. For instance, in the case of water and forests, they were both originally listed under local or state responsibilities (State List in the constitution).[3] But as the center continued to infringe on state rights, it led to an imbalance with the center formulating policies that were the states' responsibilities to implement. States were also supposed to generate their own funds to carry out these new responsibilities (Khator 1991). The Tenth Five-Year Plan (Planning Commission 2002, 5.1.46) had inadequate integration of environmental concerns, including biodiversity and resource-based livelihoods. It was only in the Eleventh Five-Year Plan (Planning Commission 2007), after the first National Environment Policy was put into place in May 2006, that there was for the first time an integrated look at environment and climate change, with the idea that "development strategy should be well complemented by policies for environmental protection and sustainability" (Planning Commission 2007, 3.53), although its emphasis remained on liberalization.

In addition to this, a variety of formal laws and policies governed the management of resources in India. Forests and wildlife are in the concurrent list of the Constitution of India, which has the provision for conservation of forest and wildlife under articles 48A and 51A (g). The main responsibility of the central government is policy and planning; implementation of activities related to forest and wildlife is with state governments. The policies of other sectors such as the National Environment Policy 2006, National Agriculture Policy 2000, National

Farmers Policy 2007 and Integrated Energy Policy Report 2006 also have an impact on sustainable resource use.

The importance of the five-year plans lies in the glimpse it gives us of the institutional logic underlying all major environmental decision-making since Indian independence. The five-year plans make it clear that environmental concerns were considered the purview of the central state and took a back seat to developmental concerns almost continuously. In addition, in spite of India's long tradition in community management of resources, the role of people and participation in policies managing resources or in conservation was constantly eroded by the central state. Public participation as an integral part of environmental management only surfaced very recently and was not pursued with any real commitment.

The legal foundation for the environmental issues in India is a hybrid of the American and the British policies; it has not been properly balanced to suit the indigenous context for efficient management. The Indian policy emphasizes the establishment of universal standards and a centralized process, but the bureaucracy is largely left unaccountable. Thus, informal regulation plays an important role in this process by which social pressures, such as negative scrutiny by the press or direct community regulatory action, are used as mechanisms of enforcement. Other mechanisms of informal regulation include "demands for compensation by community groups, social ostracism of the polluting firm's employees, the threat of physical violence, and efforts to monitor and publicize the firm's emissions/discharges" (Kathuria 2007, 404). Though these tactics do not offer an appropriate substitute for a strong environmental governance regime, they are useful at a community or regional level as they allow people agency. In a country as vast and diverse as India, a strong policing system by the government does not usually work especially because of existing power asymmetries. Thus these soft modes of regulation are often far more effective, especially in building stakeholder consensus and strengthening community regulation.

Given the challenges faced by the government and its policies in India, merely technocratic solutions and broad strategies to ecological problems remain merely blueprints, as societal conditions do not allow institutions to operate effectively. Consequently, theorists suggest greater transparency and public participation, as do an increasing number of Indian citizens (Qaiyum 1997). One important policy enshrining participation in environmental governance was the policy for the Abatement of Pollution in 1992. The policy is important because it declared the objective of integrating environmental considerations into decision-making at all levels. To achieve this, the main principles were (i) to prevent pollution at source, (ii) the adoption of the best available technology, (iii) the polluter pays principle, and (iv) public participation in decision-making (Divan and Rosencranz 2001, 36). This is one of the earliest examples of the institutionalization of participation. Its main focus however is on informing citizenry of environmental risks by periodical publishing. It also prescribes public and non-governmental roles in environmental monitoring to create strategic pressure groups that would complement the regulatory framework. The public's role in the actual formulation of the plan was not addressed (Abatement of Pollution 1992, sec. 11).

The role of participation in policy- and law-making in India has slowly emerged with the courts playing a crucial role. Through the 1970s, up until the Bhopal gas leak disaster in 1984, environmental law could not be distinguished from the general body of law. Post-1984, not only did laws become more stringent covering aspects that had until then remained unregulated (e.g. noise, hazardous waste, environmental impact assessment), but the old license regime was forced to make way for more regulatory mechanisms introduced by the courts. Participatory practices like consensus-based decision-making in village assemblies have existed in some communities for several centuries in India (Drèze and Sen 2002). However, emerging clauses of participation within policy allowed for the institutionalization of participation and control of citizens. For example the amendment to the Environmental Impact Assessment Notification of 1994 made public hearings mandatory (MoEF 1994 [amended on April 10, 1997]). This was done in order to provide a forum for the disclosure of all the information related to the proposed project to all concerned citizens and gave the space for the public, NGOs and affected communities to air their views and concerns. Citizens' initiatives, provisions and the statutory "right to information" enabled citizens to directly prosecute a polluter after studying government records and data (Divan and Rosencranz 2001, 2). Added to this, the Right to Information Act 2005 that set out to "to secure access to information under the control of public authorities, in order to promote transparency and accountability in the working of every public authority" (Right to Information Act 2005, no. 22) allowed an unprecedented level of participation that affected the environmental movement as well.

Thus, we find that citizens in post-liberalization India found themselves confronted with a loosening of institutional structures, allowing for more participation in the policy process. More and more citizens were encouraged to engage with the state on matters that affected them, and there were efforts to make the state more accountable to them. This had an impact on evolving civic consciousness and civil activism.

Linking participation, deliberation and advocacy

Recent studies on democratic governance propose, "[T]he best way to tap into the energy of society is through 'co-governance,' which involves inviting social actors to participate in the core activities of the state" (Ackerman 2004, 447). The term "participation" has a spectrum of interpretations in development and is often misused. According to Swiderska (2001, 12):

> It has been used to describe different levels of involvement, ranging from information sharing and gathering, to consultation, negotiation, shared decision-making and transfer of decision-making. To be meaningful, participation needs to be accompanied by a genuine intention to allow participants to influence decision-making. When there is no guarantee that views gathered will actually influence decisions, then what may be termed 'participation' is in fact 'consultation'.

Deliberative democracy connects consultations to more direct roles for citizens to co-govern and engage with more substantive policy reform and be assured of government responsiveness to their demands. Here equal weight is given to citizens deliberating on issues of public policy in collaboration with the government, and this is a prerequisite for the legitimate exercise of authority (Cohen and Rogers 1992). In contrast to deliberative politics, participation is seen as a fragmentation of society into disparate interests in conflict with one another.

Deliberative democracy takes participation a step further by involving citizens in consensual agreements, rather than relying on traditional expressions of political citizenship like voting. The shift underscores the growing movement away from authoritative modes in politics to more communicated, negotiated arrangements that involve a wide base of actors and individuals who confer legitimacy on the decision-making process. Participatory theory also points to the communicative element of democratic politics. However, priority is given to the negotiation of particular interests with individuals organizing themselves with respect to particular narrow agendas or interests in order to indirectly influence the decision makers (Papadopoulos and Warin 2007, 451).

Critics have noted several disparities between participation and deliberation (Papadopoulos and Warin 2007; Mutz 2006). It has been pointed out that although both theories aim to confer legitimacy on policy-making processes, they utilize different approaches. Diana Mutz (2006) for example suggests that the logic of participation, which is based on the attainment of predetermined objectives, does not fit very well with rational logic of deliberation based on dialogue between people with diverse points of view. This dialogue, which allows individuals to debate points of view different to one's own (Gutman and Thompson 2004; Knight and Johnson 1997), would not be a characteristic feature of those who are active in a participatory way. Participation generally involves people who are focused on their own interests and form a part of rather homogeneous social networks (Mutz 2006, 20–50). From this point of view, many authors have seen participatory theory as a contrast to the theory of deliberation. Skocpol and Fiorina (1999) underscored the "dark side" of participation where the reliance on extreme activism throws up obstacles to average citizen involvement. Others like Larmore (1994) point out that deliberation is sometimes unequal and leads to greater conflict. Sanders (1997) takes it a step further by emphasizing that rationality which is the ideal of deliberation is unequally distributed throughout society. The deliberative framework endorses a "bifurcated model of democracy" that separates participation from decision-making, "locating deliberation within an informal public sphere and decision-making in the formal public sphere" (Squires 2002, 134). The reconciliation of this dichotomy is precisely where this book positions itself.

Participation is an essential ingredient in the process of deepening democratic deliberation. It can represent grassroots resistance to powerful elites and neo-liberalization, consensus building, mediation or a public resolution to a conflict of interest. Formal processes of deliberation can also play a crucial role in heightening participation at many levels of society (Gastil 2000; Fung 2003; Price and Cappella 2002; Wunthow 1994).

The polarization of deliberation and participation in theory has been widely discussed. In order to fully appreciate the differences, one must recognize that the analysis of participation has been based mainly on studies of the nature of interest groups (Lichterman 2006; Cooke and Kothari 2004; Eliasoph 1998), while studies of deliberation have been through the implementation of deliberative experiments (Font and Blanco 2007; Barabas 2004; Smith and Wales 2000; Fishkin 1996). This lack of clear procedural design that would allow for the convergence of participation and deliberation leads to this theoretical polarization. The expansive agenda of reconciling deliberative democracy and participation is underscored by several ideas. There is recognition that the public will need to be channeled into activities that have the possibility of influencing a decision or changing a situation. Dryzek (1990) suggests that the groups or individuals involved in deliberation should identify some sort of decision-making procedure and that this procedure should be oriented around consensus formation. Deliberative theory, however, gives more weight to the process of deliberation rather than the outcomes. Rather than focusing on the role of public participation in improving decision-making, deliberative democrats understand public participation as an opportunity for public debate, confronting diverse points of view and informing public opinion (Parkins and Mitchell 2005). According to Eliasoph (1998) the problem is in the way participation is conceived, in that it takes place in a determined public space. This means gaining an understanding of both the procedures and the space that bind deliberative and participatory dynamics. In this view, participation is not merely an abstract and universal ideal, but rather it relies on the procedural aspect of how participation (and deliberation) takes place.

This idea of space is also linked to the distinctions between "strong" and "weak" public spheres in democratic theory. Strong public spheres are noted for public deliberation in institutions like the legislature or parliament, which are imbued with decision-making authority. Weak public spheres, on the other hand, are much more diffused, often concentrated in amorphous networks and coalitions that form around specific issues areas. They are also often characterized by deliberation without decision-making authority (Fraser 1990). Unlike Arnstein's (1969) ladder approach to public participation (Table 5.1), deliberative democrats see these weak forms of the public sphere as providing invaluable opportunities for public deliberation that can translate into important sources of influence on decision makers and the general public (Parkins and Mitchell 2005, 532).

The idea of "strong" and "weak" public spheres is further complicated by the idea of "strong" and "weak" publics (Fraser 1990). The former are "publics whose discourse encompasses both opinion-formation and decision-making" (Fraser 1990, 75). The latter are "publics whose deliberative practice consists exclusively in opinion-formation and does not also encompass decision-making" (Fraser 1990, 74). The weak public's role in the public sphere is to provide critical commentary on the authorized decision-making undertaken by the strong public. It is the extra-governmentality of the public sphere that confers legitimacy and autonomy on the advocacy within the public space and the public opinion generated by it. According to Fraser, this is further complicated by parliamentary sovereignty, which "functions as a public sphere within the state" (1990, 74). This blurs the distinction by encompassing both opinion-formation and decision-making as a represented public, deliberating on legally binding decisions. The case studies examined here extend these blurred sites of deliberation to government-led steering committees and joint parliamentary commissions that are referred to as participatory spaces. These capture the ambivalence of the categorization of strong and weak publics while using advocacy as the bridge connecting these different conceptualizations.

The roles of advocacy, deliberation and participation are highlighted, particularly in relation to their role in guarding spaces for deliberation against elite capture. The quality of public deliberation can be enhanced by conferring legitimacy on social policies through collective action in the public sphere. Civil society organizations and actors often play a critical role in publicly criticizing state policies and social alternatives. Addressing these perspectives on the role of deliberation and participation is crucial as both participation and deliberation are increasingly promoted as viable means to bridge schisms, deepen democratic practice and advance understanding among opposing factions (Fishkin et al. 2007). For instance, the case studies see a tension at the deliberative stage between the state's endorsement of "technical expertise" and the advocate's mistrust of it. A crucial balance has to be achieved in order to take advantage of the emerging spaces of deliberation – deliberative forums, working groups and joint parliamentary committees. These spaces have to be supported with disaggregated forums at the state or local level in the form of public consultations that feed into formal structures of deliberation. The inherent strength of this approach is that it provides a space for people to discuss the nuts and bolts of policies that directly affect them and be assured that their opinions are being conveyed to the highest echelons of the state. At the formulation stage these forums equalize representation while allowing for the public's demands to be converted into more technical interpretations of resource management for biodiversity and forestry that are more trusted by the state. At the implementation stage, these forums and the people involved within them can advance measures and proposals at the local level. In addition they can bring insights to local populations on the discussions and constraints at the central state level.

The constraints of achieving mutual understanding of policy among the formal discourses taking place in the formal state-led sphere of policy formation are evident in the case studies. In the presence of power asymmetries, narratives that are used to counterbalance more privileged, narrower interests become an important source of power. As we will see in the case studies, narratives of equity and participation were used to counter-balance the pro-development bias of the state. Deliberation is also encountered among narratives that mobilize the public sphere to act on particular problems. The public defense of particular policy choices that ensued through newspapers and television advertisements in the case of the Forest Rights Act clearly highlight this. Unlike standard approaches to deliberation, communicative power is not just limited to rational argumentation but also encompasses protest activities including civil disobedience. This activity, rather than being limited to changing the behavior of the state, seeks to build pressure on policymakers by building consensus in the public sphere. As a result, communicative power that is built in the informal public sphere is used to counter other influences on the bureaucratic apparatus and extended to the formal sphere of policy-making.

There exists a large body of literature on the theory of deliberative democracy. The implications of this literature and its implication for real-world policy-making, however, are far less well known. As Russell Hardin observes, "[T]here may be no larger literature in recent political philosophy larger than that on deliberative democracy"; however, "there is little instruction for the neophyte on just how such a theory works on the ground" (cited in Fung 2005, 417). Capturing these ideas requires designs that allow linkages between citizen-to-citizen deliberations on contentious issues, provides forums that converges people with dissimilar views, and assesses intended political participation as well as the factors that underlie mobilization. To this end, I identify advocacy as the crucial factor in enhancing both deliberation and participation. This is supported by field practitioners who point out, "There is some support for moving away from analyzing the differences among methods and instead focusing on how to institutionalize deliberation, expand its scale, and connect it to other democratic practices, such as advocacy and movement-building" (DDC 2008, 2).

Deliberative theory tends to favor dialogues over confrontation. However, when dialogue fails against powerful adversaries, activists may supplement or replace dialogue with strategic and cultivated confrontation, as an important and routine form of political action (Young 2003, 104). By analyzing the role of civil society organizations in policy advocacy, this study strengthens the idea that advocacy in the context of deliberative democracy enhances the democratic process. It expands the public sphere to civil society organizations (CSOs) that represent different groups in society so that they can deliberate and discuss diverse points of view about public policies that affect them, making the process participatory.

There are some of the limitations of deliberative spaces, especially those ones where structural and historical inequalities underlie the issues that are debated. These limitations are often encountered in decision-making relating to resources, where powerful elites, be they conservationists or the state, monopolize the deliberative space. One sees in the case of the NBSAP process that there is a danger that most government-led policy processes stop at tokenism where citizens may hear and be heard but they do not have the influence to make sure that their views will impact policy. The role of advocacy is particularly important then in integrating marginalized perspectives, guarding these spaces of deliberation and keeping them from capture by more dominant groups in politics.

Most deliberative theory takes a perspective that looks from within government outwards toward the public, while under-theorizing the role of civil society, or prescribing its role within government mechanisms (Hendriks 2006a). This book looks at the creation of strategic venues for deliberation by civil society. It considers various forums and venues for public consultation at the local level and the way civil society organizations and social movements manage to incorporate their interpretations garnered from these venues into government structures. This book flips the perspective from that typically found in deliberative theoretical approaches, as it looks outward from civil society towards government. It demonstrates how social change organizations – civil society organizations with social justice orientations (Chetkovich and Kunreuther 2006) – bring their views to deliberative venues (like joint working committees or joint parliamentary committees), whether through participation within them or direct action outside of them. Instead of simply focusing on the state and civil society as oppositional spheres, I outline the strategies by which the substantive goals of advocacy groups become more likely to be discussed and negotiated in a public sphere. In this way, I contextualize and locate the pathways by which the views of civil society and civil society organizations can be assimilated into the policy formulation process. In this approach, encompassing both top-down and bottom-up modes of participation, advocacy groups bring the worldviews of the marginalized who are far from the centers of power into the policy process. They use both international norms and the domestic public sphere to confer legitimacy on their narratives.

Identifying participatory spaces

The connection among political governance, identity and space was introduced to mainstream political science by March and Olsen (1995). According to them, the governance of this space, no matter what its size was, is about "affecting the frameworks within which citizens and officials act and politics occurs, and which shape the identities and institutions of civil society" (6). Using this idea as a point of departure, the discussions on social movements and NGOs can be reoriented to their activities in this political space and their role in facilitating participatory deliberation.

Civil society, defined as the space between government and citizens, plays a key role in this discussion. It denotes "the sphere of private institutions, organizations, associations, and individuals protected by, but outside the scope of state interventions" (Nash 2000, 273). Much of the analysis focuses on strategies to create spaces, make room for a diversity of voices and enable people to occupy spaces that were previously denied to them. This study conforms to the idea that the

> [u]nderstanding of the dynamics of the participation requires a more qualitative and subjective conceptualization of space. Political space, from this perspective, is not just filled up with competing interests but rather is understood as something that is created, opened, and shaped by social understandings.
>
> (Fischer 2006, 25)

To this end, two ideas are discussed in detail: first is the recognition that there are unequal social relationships and that civil society organizations can play a role in creating political spaces for people who were previously denied the opportunity to discuss policies that directly affect them; second is adding the idea of competing interests into the analysis of discursive practices where historically rooted processes and narratives have given credence and power to some ideas while dismissing others to the fringes of political discussion. This means that "the meanings that constitute a space are carried and conveyed through discourses; through the production and replication of power relations within institutional spaces, they serve as means for domination and control" (Fischer 2006, 25).

For the first point, the creation of new spaces for policy deliberation has been outlined as a meeting point between the government's "invited" and civil society's "claimed" or "invented" spaces (McGee 2004; Cornwall 2004). These "new democratic spaces" (Cornwall and Coelho 2005, 1) are intermediate, situated as they are at the interface between the state and society. In India, they are also intermediary spaces that are conduits for negotiation, information and exchange. They may be set up and provided for by the state, backed in some settings by legal or constitutional guarantees and regarded by state actors as their space into which citizens and their representatives are invited. Yet they may also be seen as spaces conquered by civil society, in their demand for inclusion.

In addition, basic resources, rights and capacities are required by participants to make full use of the political space. Cornwall (2004) points out the need to assess the work required with groups prior to their participation in a process, in order to ensure greater equity in participation. This idea extends to include improved advocacy skills to mobilize and organize outside of the policy arena of the new democratic space, as well as essential awareness raising and consensus building (Gaventa 2003; Kabeer 2005). The role of civil society organizations in this scenario is in providing support to participatory initiatives, including providing marginalized groups with access to information and material support. It also plays an important role in establishing "vertical lines" of communication, linking grassroots issues and structures with national processes (Stiefel and Wolfe 1994, 207).

As to the second point, this political space is never value neutral and is not rooted in any objective reality in procedure. Instead it is shaped by inherent and persisting power asymmetries and unequal social relationships. This means that the creation of institutionally viable spaces for participatory processes has the ability to constrain and marginalize certain groups, while enabling others. The constructions and meanings of particular spaces are conveyed through narratives and discourses. These narratives and discourses also determine the identities and meanings that persist in a societal context that are brought to the fore by discursive conflicts that are engaged with in political spaces. This leads to assessing participatory processes, identifying what kinds of actors and strategies permeate this space, to pose the question of who determines what forms of participation dominate a given space, who it is initiated by, what the people who enter these spaces bring with them, what beliefs and values underline certain discourses and what methods are chosen to strategically orient themselves to the space (Fischer 2006).

Ultimately the aim of advocacy in the policy process is not just to create spaces for the inclusion of different perspectives but also to preserve these spaces throughout the policy process. We see in the forest rights and biodiversity cases that institutional context and powerful external agents often threaten the democratic nature of policy processes especially when powerful factions dominate deliberative spaces. The role of advocacy then is to keep deliberative spaces receptive to a conveyor belt of ideas that are fed into it from disaggregated forums in the public sphere. This interplay of inclusion and exclusion which Dryzek (2000) points to is not just a call for participation but rather an active effort to create forums at several levels of the state where individual and community's subjective interpretations of policy content is actively discussed and negotiated. The case studies discuss the creation of these forums as well as the strategies used to link these micro-deliberative forums to the main deliberative space within the relevant ministries. They provide empirical examples of the conditions and practices that can influence the success or failure of deliberative empowerment

The case studies underline the process of expansion of democratic spaces within the state. These democratic spaces include deliberative forums, working groups and joint parliamentary committees that are not effective without being linked to active forms of advocacy and mass participation. There are a few studies that have carved out a space for social movements and civil disobedience in the context of the deliberative democratic ideal (Cohen and Arato 1992) or the role that demonstration and protest activity play in opposing state policies (Dryzek 2000). However, it is generally understood that most deliberative democratic theory is critical of advocacy that includes direct action like boycotts, protests and sit-ins, among others. I argue, however, that advocacy – including non-deliberative tactics, involving elements of civil disobedience – is crucial for preserving the integrity of deliberative spaces. The notion of deliberative policy-making used is one where advocacy and deliberation can be seen as a continuum rather than as activities in opposition to one another. Participation and institutional collaboration based on rational argumentation and consensus building is not effective without advocacy in the public sphere and strong grass-roots mobilization capacities. Particularly in the context of India, there is often a separation between the formal discourses and decision-making of the dominant elites and the social movements and protests limited to the public sphere that is directed at these policies and narratives. Yet, there are emerging, mediated spaces between the state and civil society which civil society actors use to transfer ideas from the public to the private or formal sphere. In doing so, they reconcile the aspect of direct action and advocacy to the reason-based, consensus-building model of deliberative democracy.

Rationale for comparative case studies

Two cases were chosen for this study: the National Biodiversity Strategy and Action Plan (NBSAP) and the Scheduled Tribes and Other Traditional Forest Dwellers (Recognition of Forest Rights) Act, 2006 (henceforth referred to as the Forest Rights Act or FRA). These were among the first environmental policies to achieve large-scale deliberation at the formulation stage. In addition to advocacy at different stages of the formulation process, both policies made serious attempts to include the voices and ideas of more disenfranchised and marginalized sections of society. The impetus for both policies grew out of different international and national windows of opportunity. The case studies compare the strengths and weaknesses of both the top-down and bottom-up initiatives that these policies are rooted in. Both relied on the utilization of a combination of public and private institutional spaces and exposed the paucity of truly deliberative spaces in the Indian policy process. However, by exploiting the spaces that did exist, they underlined the need for extra-parliamentary advocacy in bringing the agendas of marginalized sections of society and discussions on rights to the deliberative sphere. This study also alludes to the power asymmetries in policy-making without any explicit discussion but retains the idea that the inclusion of marginalized voices into deliberative frameworks "level[s] [the] discursive playing field, which in turn encourages a culture of competitive participation where the politics of dignity are played out, boundaries of caste and class transgressed, and the political power of the poor displayed" (Rao and Sanyal 2010, 163).

The focus on the NBSAP and FRA, which are both concerned with resource distribution and reclaiming cultural and historical rights, was important for reasons which are discussed in further detail in chapters four and five. Although the NBSAP is a plan as opposed to a ratified act, in contrast to the Forest Rights Act, it is in itself an indispensable step on the road to implementation of a holistic biodiversity policy. NBSAPs in many countries, including India, have generated important results, including a better understanding of the cultural and economic value of biodiversity, user rights and what is required to address threats to it. Legal gaps in implementation have been filled with national legislation, the coverage of protected areas has been considerably extended, and grassroots initiatives have sprung from the activities around formulating the NBSAP. It has also contributed to mainstreaming biodiversity within sectoral and cross-sectoral activities at the national and sub-national levels.

In contrast, the Forest Rights Act is a ratified policy that is ready to be implemented which empowers the tribal communities and people and other traditional forest dwellers to protect their access and use of forest resources. It recognizes and vests the traditional rights to forest dwelling communities over access to forest goods and occupation in forestlands, addressing centuries of colonial repression and decades of post-colonial policy that were at odds with the rights and needs of tribal and local communities.

The biodiversity and forest rights cases had ambitious agendas as they wanted to overturn systems, ideas and historically ingrained discourses in favor of people-centered, rights-oriented approaches in policy. These policies that were predicated on ideas of justice and equity came into direct conflict with the mainstream discourses and narratives of ecological modernization and conservation. In this context the two policies were groundbreaking in opening the door to participatory policy-making where citizens were encouraged to take part in the debates on policies that affected them, mobilized to push them through and directly contested policy issues and discourses in the public sphere that they deemed harmful. These processes opened the door to a more participatory style of governance which allowed for consensus-oriented decision-making, a more inclusive approach, a more participatory style of structuring policy formulation and a responsive civil society and state.

This book traces the conflict lines between the different coalitions and actors involved in both policies. The National Biodiversity Strategy and Action Plan "involved a high level of informed debate" (Anuradha et al. 2001, 22) and generated large-scale participation by civil society organizations, but it ultimately was not recognized by the state as an official plan. In contrast, the Forest Rights Act, strengthened by large-scale mobilization from its inception, was hailed as one of the "most revolutionary contributions to [the] tribal law making process in India" (Patnaik 2008, 9). It must be noted that both the NBSAP and the FRA were ultimately regarded as failures to different degrees – the NBSAP for the stonewalling it faced at the last moment by the state, and the FRA for the dilution it faced in the crucial moment before its passage in parliament, which limited the provisions negotiated and discussed over the course of the formulation process. However, both were extremely important in ushering in a consciousness in participatory policy-making. Both policies forged new strategies towards a more inclusive formulation process and carved out room in existing institutional spaces to push agendas that stood in direct contrast to mainstream state and societal discourses. In doing so these processes led by civil society organizations directly challenged the command and control system of environmental management retained by the state. It is also important to keep in mind that although India is the world's largest democracy, actual decision-making follows a representational model where the main actors are elected representatives that retain the bulk of decision-making authority and not the people themselves.

The cases can be contrasted as one that failed and one that succeeded. The rationale for studying a successful engagement with deliberation was to contribute to the understanding of the conditions under which coalitions emerge and the strategies of advocacy that succeed in interactions with the state. The examination of an unsuccessful case, as in the case of the NBSAP, is to understand the weaknesses of certain forms of collective action in spite of advocacy strategies. The study of emerging collective action allows us to capture the dynamics of what facilitates or frustrates advocacy around environmental issues including the role of the state and political opportunities. With these issues in mind, I chose the arrangement of coalitions, with specific advocacy strategies and narratives, around the National Biodiversity Strategy and Action Plan and the Forest Rights Act.

In the National Biodiversity Strategy and Action Plan case, the two main coalitions identified were the actors clustered around the *pro-participation coalition* (technical and policy core group). These eco-populists advocated for greater inclusion of marginalized voices whose narrative I have referred to as "civic environmentalism" in my case study. This has been contrasted with the proponents of "ecological modernization" who were actors in and around state institutions who pushed for the policy to be kept within state control with participation limited to only state-approved actors. In the case of the Forests Rights Act, these coalitions were supplemented with additional narrative conflicts between conservationists or the *wildlife lobby* (who favored a command-and-control environmentalist discourse as their conception of rights extended to both humans and animals) and the tribal communities or the *forest rights lobby* (who pushed for greater justice and equity in the law) and the actors in-between – the *developmentalists* (who pushed for an opening up of the system to diverse voices but were also sympathetic to the environmentalist discourse). In popular parlance the narrative contention between the two main groups came to be publicly referred to as *tiger versus tribal*.

Table 1.1 simplifies the complexity of the arguments in the case studies and distills them to certain clear narratives that are common to both cases. Each coalition has a constellation of typical institutions or organizations, which advocate for a specific certain narrative storyline that is a tool of political advocacy. Through these narratives, the actors impose their own sense of a coherent policy argument.

Table 1.1 Delineating coalitions (adapted from Wittmer and Birner 2005, 4)

	Conservationist	*Eco-populist*	*Developmentalist*
Typical proponents	Conservation NGOs, biologists, ecologists	Advocacy NGOs, cultural anthropologists	Development organizations (state, NGOs, donors), economists
Central storylines	A minimum area of undisturbed nature needs to be preserved to avoid species loss and to maintain the ecological balance	Local/indigenous communities are the only true stewards of the environment. They have proven that they can preserve forest resources better than the state	Population increase and poverty are the main causes of deforestation and biodiversity loss; poverty reduction is essential for saving the environment
Priorities / mission	Nature conservation, protection of endangered species	Empowerment of local people and communities	Poverty alleviation, state lead development
Self-positioning	Defendants of nature and endangered species	Defendants of indigenous rights	Defending state role and vision as protectors of the poor
Positioning of opponents (other representation)	Local people seen as eroding natural resources; eco-populist NGOs seen as neglecting ecological realities; state and private sector seen as taking advantage local communities	Conservationists seen as neglecting human rights; state/private sector seen as taking advantage of local communities	Eco-populists seen as romanticizing and instrumentalizing local people; conservationists seen as neglecting the need for poverty alleviation and hindering development for the larger society
Relation to science	Results of natural sciences (conservation biology, ecology, hydrology etc.) as unquestionable basis for the argumentation	Postmodern criticism of science; reliance on qualitative social science studies and on natural science studies challenging "orthodoxies"; high valuation of local knowledge	Reliance on technical disciplines (agronomy, engineering etc.) and on socio-economic studies

They position themselves clearly as "defendants," "victims" or "heroes" who are struggling to preserve a certain vision or way of life.

Though the scales of the two policies differ, both are examples of the way policy meanings are shaped in contested deliberative environments. In their formulation, both contested issues about redistribution and provided spaces where organizations responded to the state. They both confronted overarching mainstream narratives by reframing issues based on local experiences or global norms. Both attempted to redress long-standing inequalities stemming from failed public policies. In addition, as environment policy is often very scientific, technological (Fischer 2000) and expert driven, these cases represent very effectively different narrative conflicts between "scientific" and "local" or "indigenous" knowledge. This context allowed me to highlight controversies over meaning and also over historical and contemporary narratives of resource use and participation. The two case studies also demonstrate path dependencies and long-entrenched interest groups and paradigms, which give way to a multitude of contentious narratives in the contemporary policy space. Finally, I contrast the advocacy by actors to push for more participative forms of deliberation in policy formulation, specifically their strategies in both formal, state-led spaces and public, informal spaces. The "invited spaces" were applicable to the case of the NBSAP, which was initially conceived by the government as a formulation process in collaboration with civil society. The "invented spaces" were applicable to the case of the FRA that claimed its space within institutional structures. Using these cases, I am able to highlight advocacy strategies utilized in the informal sphere that are necessary for successful negotiation in policy formulation within formal institutional structures.

While international scholars have lauded participation, few questions have been posed as to what the participation of people in decision making actually entails. Radical scholars have pointed out that participation is either tyrannical (in the hands of the powerful) or transformative for marginalized segments of the population (Cooke and Kothari 2004; Hickey and Mohan 2004). There has been a growing group of scholars who identify that "[p]ublic deliberation is essential to democracy" (Page 1996, 1). The emergent theoretical idea of participatory spaces is a recent development pursued by a few scholars who have traced the evolution and shaping of these spaces by local agents. Although there are multiple studies on actors, ideas, narratives etc. in public policy-making, very few studies have applied Cornwall's (2004) theory of participatory spaces to a large-scale deliberation in which multiple agents, representing various sets of interests, are attempting to influence a policy process. This enables one to see the shifting character of both agents (who create participatory spaces) and the government or institutional structure (which responds to particular demands). More specifically, these case studies ground those responses in historical and contemporary narratives that shape the activities within participatory spaces. In addition, they underline the conflict in narratives and the acceptance of certain interests within such spaces. These cases of rights over biodiversity and forests are revealing of the changing nature of people's participation and responses in participatory spaces given the changing structures and institutions of the state and civil society.

To this end, these cases present an answer to an intriguing puzzle – why did two policies with similar characteristics achieve such disparate results? Both are post- liberalization policies, both involve a large number of resource-dependent livelihood communities and both deal with the complexities of resource use and mobilization. Both biodiversity and forest rights are organic concepts that did not exist in the lexicon of environmental governance in India before 1999 and 2006, respectively. Both involved a large number of actors with a diverse range of institutions, both within the state and in civil society.

In order to understand why some processes fail while others succeed, comparisons between similar types of projects with different outcomes are necessary. Policies even within the same policy space have heterogeneous characteristics, different dynamics shaping demands, different responses to international pressures and norms and different coalitions of interest groups. Thus, in this book, I seek to unravel the puzzle of different outcomes by comparing cases which suggest that while the model of deliberative democracy is relevant in contested terrain, there are additional extra-parliamentary strategies that have to be married with the model in order to see policies to fruition in complex environments.

Methodology

In order to understand the advocacy strategies of civil society within deliberative democracy, I used elements of an inductive, interpretive study in the form of narratives to arrive at rich descriptions of deliberative democracy in the cases of biodiversity policy and the Forest Rights Act of India. Qualitative research is a process of understanding a social or a human problem based on distinct methodological traditions of inquiry. The researcher "builds a complex, holistic picture, analyzes words, reports detailed views of informants and conducts the study in a natural setting" (Cresswell 1998, 15). This idea is supplemented by Denzin and Lincoln (2000), who claim that qualitative research involves an interpretive and naturalistic approach: "This means that qualitative researchers study things in their natural settings, attempting to make sense of, or to interpret, phenomena in terms of the meanings people bring to them" (3). In contrast to positivism, which nurtures nomothetic ambitions to give generalized explanations for cases, interpretive policy analysis offers ideographic explanations that are limited to particular, individualized contexts. The qualitative ideal is not a generalized law, which governs the events independently of particular place and time contexts, but plausible thick description of the actors' understandings and interpretations, the presence and role of narratives, discourses and arguments in a social context. As Lincoln and Guba (1985, 110) suggest, "[R]ealities are whole [and] cannot be understood in isolation from their contexts" (39), and they are constructed, holistic and multiple.

Both the conceptual framework and the research questions reflect mechanisms through which deliberative democracy draws on values, beliefs, emotions and experiences, which add to new information through which conflict can be resolved and consensus reached. This conceptualization draws upon the work of pragmatists like Charles Sanders Pierce (1877), John Dewey (1929) and William James (1902), who understood the truth as being contingent on beliefs.[4] This approach makes space for the relative influence of the situational context, the socially constructed reality, linguistic cues and social relations. In addition, it puts particular emphasis on the idea that it is impossible to understand an institution adequately without an understanding of the historical process in which it was produced (Berger and Luckmann1967, 72). By focusing on the construction, meanings and influences of participant narratives within the case studies, the analysis explores the way deliberation constructs meaning and guides policy shifts.

The turn of policy analysis to interpretation, analysis of narratives, storylines and discourses has several mutually interconnected theoretical underpinnings. Theorists in this framework view policy issues as both symbolic and substantive, and believe that they can be constructed in different ways. These constructions are often in competition or opposition to each other and convey a story or myth about why it is a problem to begin with, who is rewarded or disadvantaged by the problem, and how the problem can be solved (Birkland 1997; Schneider and Ingram 1993; Yanow 1992, 1996). The viewing of a policy through a positive lens, in contrast, claims access to an "objective" world. Mainstream positivist frameworks neglect the "fact that the very same policy is a symbolic entity, the meaning of which is determined by its relationship to the particular situation, social system, and ideological framework of which it is a part" (Fischer 2003, 60). Stone (1988) defines one type of social construction as causal stories, which are human-centered and constructed by political actors using language to make the issue fit their position while trying to capture the largest public interest (Birkland 1997; McBeth et al. 2007). Stone's work uses literary devices – including plots, heroes and villains, and metaphors – to scrutinize policy narratives (McBeth et al. 2007). Yanow (1992) focuses on myths created by policy narratives, which are created and believed by different groups of individuals in an effort to divert the public's attention away "from a puzzling part of their reality" (401). Deborah Stone (1998) sets forth the argument that policy analysis and any other action that attempts to interpret relations in a political community is in principle a political act. Stone (2002) understands political community as a polis, governed by "laws of passion," not by "laws of the market," in which communities and conflict groups come into conflict but also process and enact different political narratives.

In the allocation of these roles, defining interests and creating analysis, actors often reclaim narrative stories and metaphors within it. They draw on collective ideas and tacit knowledge to create a context for action that gives credibility to policy choices and demands. I trace the historical roots of certain positions taken by the different actors in the policy formulation process pertaining to the two policies. I engage with the narratives and metaphors, which strengthened different points of view. To this end, I explore the link between perception and action, moving from description of the actors' positions to the way these narratives influence political problem definition.

Interpretive policy analysis uses qualitative methods. In its broadest sense, this means a small number of cases, or even just one case, are intensely researched with thick description, often in the form of a narrative. Narrative enquiry has many elements common to case studies: they both focus on the context of embedded phenomena, case studies often report activities in narratives forms, and they aim to get as close to practitioners' points of view as they can (Clandinin and Connelly 2000; Ospina and Dodge 2005; Yin 2003). While narratives capture the practice of the political community, case study methodology is useful in providing a comparative logic.

In order to understand the advocacy strategies of civil society to increase the diversity of voices within deliberative democracy, I used elements of an inductive, interpretive study in the form of narratives to arrive at rich descriptions. This allowed me to capture the role of advocacy in deliberative democracy in relation to biodiversity policy and the Forest Rights Act in India. This was combined with comparative case study analysis. As Eisenhardt and Graebner (2007, 25) explain:

> Multiple cases are discrete experiments that serve as replications, contrasts and extensions to an emerging theory. But while laboratory experiments isolate the phenomena from their context, case studies emphasize the rich, real world context in which the phenomena occur.

The case selection was guided not by theory but because the National Biodiversity Strategy and Action Plan and the Forest Rights Act are among the first examples of large-scale participatory processes in India. For the first time, nascent spaces of deliberation opened up within institutional structures that encouraged participation of ordinary citizens and civil and social movement organizations at many levels of the public and private sphere. I chose these cases in order to illustrate the difference in outcome for the two case studies: both of which address the changing nature of people's participation and responses in participatory spaces, given the changing structural circumstances that are occurring nationally in India and at the international level. This book contributes to a nuanced rebuttal to the idea that "support from the society as input for decision-making is less significant in the developing country context" (Osman 2002, 39). State-society relationships are being interrogated world over, especially in the context of applying deliberative governance frameworks. As there have not been any in-depth studies that analyze how advocacy strategies from the public sphere push agendas of participation and inclusion into the deliberative sphere, this study creates a contextualized understanding of state-society relationships in a developing-country context.

The first stage of this research was the identification of a hypothesis which was that state-led institutions were more likely to successfully deliberate on issues if civil society organizations engage in long-term strategic and extra-parliamentary advocacy. I set out to empirically explore how and why civil society organizations use advocacy to exploit policy windows and push participative forms of deliberation in policy formulation. The key questions posed are as follows:

1   How do civil society organizations respond to different policy openings in invited spaces (state-led spaces of participation) and in invented spaces (community-led spaces of participation) (Corwall 2002)?
2   Under what conditions have participatory spaces emerged?
3   What are the narratives at function and in conflict within these spaces?
4   What does this imply for the practice and theory of deliberative democracy?

This process relies upon a cyclical process of linking extant theory, emergent data and adaptive theory. Given the exploratory nature of the research questions, the intention was to observe interaction processes in a naturalistic setting, with particular emphasis on how the broader community and historical context influence interactions within a political process (Yin 2003), and the strategy employed for this research is the comparative case study.

It was also necessary to define "policy formulation" as the key focus of this study. In addition, the emphasis of this book is on the "process" of policy formulation rather than the policies' contents or implementation. It has been argued that implementation creates a feedback loop, by which policy formulation impacts and influences implementation. However, in light of the view that implementation is "an ongoing process of decision-making by a variety of actors" (Grindle 1980, 5), these feedback loops are less common in developing countries (Osman 2002). This supports the clear demarcation that this study makes. The policy formulation stage is conceptualized as a particular arena, in line with the ideas of Jos Mooij and Veronica de Vos (2003), who point out that every arena is a specific set-up of institutions, actors and stakes. They go on to explain that particular sets of institutions and actors who may be influential in the policy formulation arena may be marginalized in others.

The second stage of my study identified cases that have had large-scale participation, where there have been spaces for deliberation and conflict that resulted in advocacy by the participants and contained explicit tensions in key narratives. Although both policies shared all these characteristics, there was a discrepancy in outcome with the National Biodiversity Strategy and Action Plan failing to be accepted as an official policy document while the Forest Rights Act was successfully ratified into law. To analyze this discrepancy I built a theoretical model around the idea of Cornwall's (2002) "invented" and "invited spaces" in order to better understand the actor constellations and strategies around these deliberative spaces. The third stage integrated the findings of my fieldwork into the proposed model to analyze the conclusions. Although my explanation of methodological strategy has been structured into different stages, this study was an iterative process, which used an adaptive approach where themes that emerged from my empirical findings were tested against pre-existing concepts and theories resulting in a fuller identification of key processes.

I use a qualitative research paradigm. The general idea of this type of research is to "understand experience as nearly as possible as its participants feel it or live it" (Sherman and Webb 1988, 7). Some of the characteristics of qualitative approaches are that studies take place in their natural setting; multiple interpretive methods are utilized; the process is iterative; social phenomena are studied holistically; and researchers fulfill the role of interpreters (Bogdan and Biklen 1992; Creswell 2003). Merriam (1998, 39) discusses combining interpretive methods to case study research and underlines:

> The level of abstraction and conceptualization in interpretive case studies may range from suggesting relationships among variables to constructing theory. The model of analysis is inductive. Because of the greater amount

of analysis in interpretive case studies, some sources label these case studies analytical. Analytical case studies are differentiated from straightforward descriptive studies by their complexity, depth and theoretical orientation.

A qualitative approach was adopted as it can offer "valuable insights into how people construct meaning in various social settings" (Neuman 2006, 308). The construction of meaning is particularly important in testing the conflicts in narratives propagated by different actor constellations in the policy space. Yanow (1995) points out that the study of built spaces is important in interpretive policy analysis. She explains that settings of policy and agency may communicate policy meanings other than, in addition to, or even contradicting those named in the policies themselves. While the case study is useful in comparing the comparative logic of the two policy processes, narrative enquiry allows one to analyze practice and delineate the discursive conflicts through the use of policy narratives. Policy narratives are particularly relevant to the study of deliberation as it "constructs a relational form of deliberation in which participants appeal to common values and experiences through telling stories. In this manner, narrative supports a form of deliberation that stresses equality, respect for difference, participation and community" (Ryfe 2002, 360). In particular, analyzing the structures and content of the stories told by policy actors facilitates comparisons between the different versions of stories, scenarios and arguments and allows us a more holistic view of the different directions policies can take.

These methods discussed above are validated by expert interviews with key researchers and policymakers and supplemented by content analysis of documents from the policy process, including discussion papers, newspaper articles, background texts and formal legal documents. The qualitative case study method is best suited as it is effective in locating the "perceptions, assumptions, prejudgments, presuppositions" (Van Manen 1977), and for connecting these meanings to the social world around them. Detection of actor constellations around certain discourses was achieved through "reputational sampling techniques" in which known actors are asked to identify others, or by identifying participants in relevant policy processes.

My fieldwork was completed in two phases in India (May–July 2009; March–June 2010). I have conducted thirty-two semi-structured expert interviews and collected material to review the entire processes. My methodology has followed a combination of semi-structured interviews and identification of narratives in order to get more representative views on the structure and debates within the policy process. Data collection included publicly available information on actors working in different public and private institutions that are relevant to the formulation of the policies, semi-structured in-depth interviews, communication material between the actors involved in the process, informal discussions with members of the wider community and policy reports, newspaper and magazine articles, papers and briefs.

The analysis for this study was conducted in three phases. The first phase took a broader historical view of the context within which the two policies were embedded.

Secondary data, namely historical and documented records, provided evidence for the retelling of the forest policy history and the durability of the coalitions and constellations within the policy area. My analysis focused mainly on creating "a storied account of practice" within the cases (Clandinin and Connelly 2000) and making comparisons across the cases. This meant analyzing the narratives. For transcript analysis, I employed Boyatzis' (1998) method of thematic analysis and code development. Thematic analysis is "a process for encoding qualitative information" and "requires an explicit 'code'" (Boyatzis 1998, 4). I began by reducing each interview transcript into an outline of two to three pages. I examined the transcripts for references to any type of narratives policymakers are using to explain their actions within the two policy processes. Some of the codes included understanding of participatory space, defining rights, conflicts between technical and traditional knowledge, advocacy strategies, importance of participation in formulation, understanding of conflicts, roles of advocacy groups and critical junctures.

The material also focused on the micro-dynamics/constructions that comprise these narratives. For example, various perspectives of each participant regarding the conflicts in formulating the two policies as the use of narrative come from different points of view/perceptions. This was also supplemented by reading the way policies were written about in workshops and in the media, as well as published papers that all focused on specific narratives the policy were built on. Some recurring tropes were "independent management of biodiversity" and "mass participation" in relation to the biodiversity plan and the "historical injustice" and "tiger versus tribal" in the Forest Rights Act. More specifically, the transcribed interviews were searched and highlighted every time the interviewees used "rights" and other forms of the word. From this information, categories were created as to how "rights" was used by all of the interviewees. By the juxtaposition of narrative conflict in my coding, I was able to position actors within certain constellations around each policy. This was also supplemented by looking at their published works and public and private interviews.

Boundaries of the study

Having outlined the key ideas, aims, rationale and approaches of this book, this section will clarify the boundaries of this research by pointing to some issues that fall beyond the scope of this study. First, this book does not seek to test how the ideal theoretical models of deliberative democracy play out empirically. Instead it seeks to evaluate the interaction of actors and narratives in emerging spaces of institutional deliberation. It concentrates on state-led forums and arenas and captures the strategies of civil society organizations in pushing agendas into those spaces and opening up these spaces to a diversity of voices. This is in keeping with the ideas of Innes and Booher (2003, 55), who explain, "[T]he tension between cooperation and competition and between advocacy and inquiry are the essence of collaborative policy making."

Second, although this study is focused on deliberative practice, participation and advocacy by civil society actors, there are several empirical issues that fall outside its scope. Most notably, I do not study the impact of deliberative designs on citizen or individual participants and the corresponding shift in preferences, empowerment and civic consciousness. Instead my focus is on civil society organizations and their indirect impacts on more inclusive, rights-oriented policies. In addition I look at the strategies and conflicts that arise from advocacy in the public sphere and its impact on the private sphere of experts, government officials and narratives.

Third, the starting point of this study was that it is widely recognized that citizens have increasing opportunities to deliberate in a wide variety of settings: in town meetings of various kinds; local, state and regional boards and commissions; hearings that solicit citizen testimonials; workplaces, civic groups, and activist groups (Crosby 1995; Eliasoph 1998; Fishkin 1996; Gastil 1993; Gastil and Dillard 1999; Williams and Matheny 1995; among others). In analyzing the spaces that are most conducive to large-scale civil society engagement, I do not outline the deliberation in small groups or particular deliberative designs. Instead my focus lies in participatory governance practices that create intermediary spaces that readjust the boundaries between the state and its citizens (Cornwall 2002).

Fourth, Fung and Wright (2003) isolate a set of characteristics that they define as "empowered deliberative democracy" that seeks to deepen the abilities of ordinary citizens to effectively participate in the shaping of programs and policies relevant to their own lives. They focus on five deliberative experiments that promote active political involvement of the citizenry. This book, rather than focusing on the experiments of empowered democracy, seeks to supplement the ideas by analyzing deeper political, historical and subjective factors that affect participation in deliberation. It has been pointed out that "the kinds of design structures and procedures suggested by Fung and Wright can offer an opening for participatory empowerment, but they cannot ensure such participation itself" (Fischer 2006, 24).

Finally, this study does not touch upon structure of deliberation in terms of its emphasis on rationality and an "objective truth" or formal procedures of unanimity and majority rule. While it is clear that the interaction of these features can make a big difference, both to the outcome and to the process of deliberation (Mansbridge 1983; Kameda 1991), these issues lie beyond the scope of this book. This book's focus on social movements, advocacy and discursive practices guides the analysis from structural and procedural deliberative practices to exposing underlying and implicit assumptions about complex social and political realities. In particular it emphasizes the role civil society organizations play in organizing and constituting spaces for participation.

Summary and organization

This book is composed of five chapters. Following this introductory chapter, the next chapter outlines the theoretical model that forms the basis of the argument. It articulates the different aspects of deliberation that the book engages with. Furthermore it sets out the theoretical model integrating aspects of both the "bottom-up" and the "top-down" policy process. Following the contours laid out in the conceptual framework, chapters three and four take an in-depth look at the case studies. The former explores the process by which the Convention on Biological Diversity (CBD) created windows of opportunity that civil society organizations exploited in order to foster policy deliberation in the creation of the National Biodiversity Strategy and Action Plan for India and the underlying reasons why the participative process was a failure; the latter analyzes the successful deliberative and advocacy process around the Forest Rights Act. It discusses the process by which civil society organizations highlight conflicting discursive frames around which mobilization occurs. It traces the mobilization of communities, who respond collectively to build strategies from the top-down to make space for rights-based claims within the deliberative space, struggling against traditional command-and-control discourse of the state and exclusionary politics. Chapter five answers the broad questions set out in the beginning of this book, highlighting points of similarity and difference in two cases. It gives broader implications for the theory and practice of deliberative democracy and what it means for real world policy.

Notes

1 This is particularly interesting because although conservationists and the government are traditionally at odds with each other, they formed an alliance in the face of the Forest Rights Act, which was perceived as a threat to both their interests.
2 The Green Revolution was a series of research and technological transfers involving improved methods of irrigation, the introduction of high-yielding varieties of seed and better irrigation facilities. The first wave of the Green Revolution started in the late 1960s, with which India attained food self-sufficiency within a decade by the end of the 1970s. However, it was limited to wheat in northern India. The second wave of the Green Revolution was in the 1980s, which included rice, covered the entire country and contributed to raising rural income and alleviating poverty.
3 Subjects defined and enlisted under the List – II of the Seventh Schedule of the Constitution of India contain subjects that form the exclusive domain of each one of the State Governments within India. The Central (Union) Government has no jurisdiction in framing laws under these subjects.
4 The idea central to pragmatism, also underlining this study, is that knowledge claims arise out of actions, situations and consequences rather than pre-existing conditions (as in positivism) and is concerned with applications and solutions to problems (Creswell 2003).

2 Advocacy in deliberative democracy

Participation in environmental planning processes has its roots in an incremental opening up of political institutions, and it is also linked to windows of opportunity in both the international and the national policy spaces, which is discussed in further detail in the case studies. The power to participate politically is given only incrementally by the state and often has to be captured by social movements or civil society organizations. Thus, both movements and advocacy by civil society organizations play an important role in carving out participatory niches in the rigid structures of decision-making and legislation. This chapter outlines the theoretical framework that I will use to analyze particular variables that influence participatory policy formulation in the context of India's two environment policies: the Forests Rights Act (2006) and the National Biodiversity Strategy and Action Plan (2009). Existing government decision-making structures are set in top-down policy processes, and it is only possible to analyze certain sectors where the state has allowed a more nuanced approach to governance.

I contend that advocacy by civil society organizations (CSOs) plays a role both within and beyond the deliberative spaces that the Indian state provides. According to Rektor (2002, 3), "[T]he history of advocacy parallels the development of democratic societies." Social policy advocacy is a form of civic participation that is also a form of citizen involvement and engagement as well as citizen action and civicness. Panitch (1993) and Abelson et al. (2003) demonstrate that there is an increasing interest in democratizing public policy processes through an engaged and informed citizenry. Civic action and advocacy in India have been active in the spheres outside the state, where advocacy encompassed forms of protests like demonstrations, *rasta rok* (road blocks) and *hartals* (strikes) because of the non-participatory nature of government decision-making. However, with nascent spaces of deliberation opening up, we can begin to identify the changing strategies and profiles of advocacy.

Dobson (2003) lists several benefits to civic participation within democratic societies: civil society actors and organizations offer experiential or traditional knowledge that can help solve local problems; the greater the diversity of people, the greater the chances of effective solutions; public participation helps reduce exclusion of marginalized groups; citizen involvement ensures political legitimacy; and advocacy enhances democracy and increases social capital as people gather to work on common problems. In this research I contend that in order to negotiate within a deliberative space and arrive at a more participatory policy process, civil society actors must engage in extra-parliamentary forms of strategic advocacy.

The concept of advocacy has an interesting history. The word advocacy originated in the legal field and was later appropriated in the civil rights movements of the 1960s and 1970s in the United States. The use of this term expanded in three key directions, all of which were linked to the concept of justice. It was initially used in relation to defending the interests of excluded groups, to refer to a proactive public interest strategy in order to change a number of established rules, as well as to defend against the abuse of public power (Fox 2001). As such, advocacy has been considered the core activity of CSOs active in the public sphere (Andrews and Edwards 2004; Salamon 2002). In India, advocacy can be seen as rooted in three different historically specific impulses: first, in the nationalist struggle for the Indian independence; second, in opposition to the largely top-down processes of governance that have defined the political culture and social systems prevailing in the country for the last fifty years; and, third, in the prevailing practices of civil society and non-party political formations. Samuel (1989; quoted in Kumar 2006, 1) defines advocacy in the Indian context as "[a] planned and organized set of actions to effectively influence public policies and to get them implemented in a way that would empower the marginalized."

Most scientific studies in the field of advocacy have originated in the United States and Europe and have become influential in a number of countries. Yet in the case of India, despite the considerable attention social movements – such as anti-colonial movements, environment movements and the like – have received, the concept of advocacy has been seldom discussed, and few studies on the subject have been published.

The terms NGOs, grass-roots organizations, community-based organizations and civil society organizations overlap. There is no clarity as to whether these terms cover organizations that operate only at the local level or also include local branches of national organizations. Grassroots and community organizations clearly refer exclusively to the local level, but civil society can denote groups at any level within a single country, and even at a global level (Eade and Ligteringen 2001, 27). "Civil society is historically an evolved form of society that presupposes the existence of a space in which individuals and their associations compete with each other in the pursuit of their values" (Clayton 1996, 39). This space relates to

> the part of organized social life that is open, self-generating, and autonomous from the state, bound by a legal order of a particular nation. It involves citizens acting collectively in a public sphere to express their interests, passions, preferences, and ideas, to exchange information, to achieve collective goals, to make demands on the state, to improve the structure and functioning of the state, and to hold state officials accountable.
>
> (Diamond 1999, 221)

Non-governmental organizations, or NGOs, constitute a "diverse group of formal organizations that are self-governing and non-profit, have some degree of voluntarism, and are expected to produce a public benefit" (Kramer 1998, 6). Several terms are used in lieu of NGO and these include private voluntary organization (PVO), private development organization (PDO), civil society organization (CSO), community-based organization (CBO) and environment and development organization (EDO). For the purposes of this study, we will be using civil society as the larger space of competing dynamics. NGOs will refer to specific organizations that are embedded within particular processes of governance but retain their independence in terms of values and sphere of influence; CSOs will refer to both NGOs and grassroots organizations that work in tandem and are more marginalized with respect to the state but with a larger sphere of influence within civil society.

Advocacy in deliberation

Deliberative democracy consists of multiple participants representing assorted interests (Weber 2003). Deliberative theory "begins with a turning away from liberal individualist or economic understandings of democracy and toward a view anchored in conceptions of accountability and discussion" (Chambers 2003, 308). Discussion is a key idea picked up in this study where the "deliberative process" is characterized by one that allows the receiving and exchanging of information between groups of actors, who can come to a consensus after critically analyzing the issues set before them. This ultimately informs the decision-making process and has the ability to broaden the scope and perspective of all participants. Deliberation's consensus-based decision-making facilitates the incorporation of ecological values in the policy-making process, allowing for more environmentally sound decisions to be made (Dryzek 2000; Smith 2003; Weber 2003). In many respects, deliberative groups and institutions are more able than traditional institutions to deal with complex, uncertain collective-action problems that plague many environmental issues.

Deliberative democratic theory rests on streamlining a political process, not through the majority or force but through reason. Central to the idea of reason and rationality are its links with public dialogue through which reason is communicated. Real democracy, the theory argues, is rooted both to communication and to deliberation, which are both institutionalized. Proponents of deliberative democracy argue that there are two spheres of power in democratic politics. One is the power created by authority and resources, usually within state limits, and the second is created in the communicative arenas of civil society and the public sphere. Theorists argue that this "communicative power" (Arendt 1970, 40) comes into play both in the public space and in the parliament. This communicative aspect is revisited in many definitions of advocacy as well (Harvie 2002; quoted in Wight-Felske 2003, 324). Rektor (2002, 1), in his definition, points to "the act of speaking or of disseminating information intended to influence individual behavior or opinion, corporate conduct, public policy and law." One finds that in fact, the concepts of advocacy, voice and social justice are inextricably linked, rooted in the epistemological structure of deliberative democratic theory. Voice, language and reason urges the theory's actors to test their world views and ideas, argue them and finally choose a mode of resolution that is the most rational. However, theory has kept the direct action associated with advocacy very much in opposition to rationality of deliberation.

One mode of reconciling deliberation and advocacy is through the use of policy narratives. Environmental policy-making occurs in highly contested and complex policy spaces that have several actors who interpret or highlight issues, de-select others and frame narratives in specific ways in order to construct a persuasive and consistent story. In a participatory context, environment policy requires consensus. Narrative policy analysis investigates the "issue oriented stories told by policy actors, using analysis to clarify policy positions and perhaps mediates among them" (Yanow 2000, 58). Analyzing the structures and content of the stories told by policy actors allows us to compare opposing elements, structure and versions and offers us a more holistic view of the different directions policies can take. "Treating story narratives as a metaphor, the policy analysts can be lead to identify 'protagonists' and 'antagonists' in a policy actor's story about an issue, the metaphors that describe the relationships between them, and the anticipated or desired transformations in them or in the policy situation captured by the plots conflict or tensions and resolutions" (Yanow 2000, 58, 59). Even when communities and groups come into conflict with each other, actors enact and disseminate different political narratives in order to make the issue fit their position to capture the largest public interest. A number of theorists in public policy and governance indicate that policy narratives are an important aspect of political advocacy (e.g. Kaplan 1986; Roe 1994; Hajer 1995; Stone 2002; Fischer 2003). Moreover, a growing body of empirical literature has demonstrated the impact that these narratives have in applied policy settings (e.g. Bridgman and Barry 2002; Bedsworth, Lowenthal, et al. 2004; McBeth, Shanahan, et al. 2010).

While discussing with the overlaps and linkages between advocacy and deliberative democracy, one also grapples with the idea of representative democracy and legitimacy. Democratic deliberation does structure and shape the political process. Some (Elster 1997) have seen it as an alternative to bargaining between competing interests – considering deliberation an active, public political act. However, it is unrealistic to believe that democratic deliberation should somehow embody the essential democratic principles of responsiveness to public wishes and the political equality of every member of that public. Habermas (1996, 1987) deconstructs this idea in his exploration of extra- parliamentary forms of representation, particularly new social movements and other kinds of civil society associations. This fits into Ezell's (2001) notion of advocacy that consists of "those purposive efforts to change specific existing or proposed policies or practices *on behalf of* or with a specific client or groups of clients" (23; emphasis added).

This is especially relevant to the political process in India because of its historical top-down approach to public policy implementation. Traditionally, environmental policies are enacted by bureaucrats who are given what are often vague policy mandates and then use their positions of power and expertise to devise implementation strategies (Birkland 2005). Deliberation, as envisioned by theorists, is not a common activity for bureaucracies, who may consult or collaborate with the public, but not deliberate with them. For example, there are comment periods or public forums for the public to express its thoughts about particular issues. However, these interactions typically do not involve an interactive discussion or have unconstrained dialogue that is required for deliberative democracy. That is why advocacy as representing the voice of marginalized people within the system of deliberation has a particularly important role to play as it addresses the issues of people who are powerless in their relationship with the state.

Warren (2001) argues deliberation enters at a late stage in the life of a democracy, in which potential conflicts rooted in marginalization and power differences have already entered the public discourse and are being worked out. This is true to some extent in India where political conflicts are only now entering a public sphere and the resolution is far more contentious than polite deliberation. Though rational, the tenets of deliberative democracy – (1) a free public domain for citizen debate and discussion; (2) a set of "rules" to ensure fair, equal and impartial deliberation for all citizens; (3) deliberation that is discursive, rational and dedicated to the public good; and (4) a government that uses this consensus based decision-making to create laws and policies (Gabardi 2001, 551) – are not fully in place. The public discussing political issues often requires a structure that is broader than the one deliberative democracy puts forth. As a result the highlighted policy agendas are often presented to the public through non-deliberative means. Thus, in order to transform a private deliberative setting to one which can claim democratic legitimacy, two additional requirements need to be met: (a) deliberative bodies need to represent public judgment and advocate for the broader values a society believes in, and (b) these bodies have to address and potentially include all the citizens or stakeholders to whom their collective decisions apply.

Representative politics has the potential to unify and connect the plural forms of association within civil society: in part by expanding the visions of citizens beyond their immediate attachments, but also by provoking citizens to reflect on collective futures (Hegel 1967). Modern societies are grounded in bureaucratic concentrations of power, and their scale and complexity dictate that citizens are mostly passive, mobilized periodically by elections (Bobbio 1987; Sartori 1987; Zolo 1992; Manin 1997). This portrayal of passivity of citizens is all too true in India where there is no legal right or any pre-existing mechanism in India for consultation with all the stakeholders affected by a particular law. This is supported by the view that ordinary people are not incapable of deliberation, but existing liberal democratic structures do not allow them the chance to deliberate (Pateman 1970; Peters 1999). Within the newly opened up policy spaces in the environmental policy process, representation through advocacy and mobilization of the "represented" opens a window on deliberation beyond the standard account.

The case of advocacy of collective causes goes beyond existing models of representation in analyzing the role of the advocate or activist. State-society linkages have undergone a process of transformation and deconstruction, and scholars remain uncertain as to whether new modes of representation are replacing the old (Hagopian 1998; Friedman and Hochstetler 2000; Chalmers et al. 1997). The traditional notion of an advocate is one who is chosen by the people, acts on their behalf and follows a precise set of instructions, such as legislative representatives. In the last decades, a new concept has emerged where the role of advocacy, both within the nation state and beyond it, is to defend actors who did not appoint them to that specific role and who are not bound by any authorization. The advocate's representation grows from a sense of identification. As Cicero (1942) explains, in his description of the role of the procurator, identification with a cause becomes more important than the explicit authorization to represent the cause.

One of the critiques of deliberative democracy comes from Chantal Mouffe (2000b). She criticizes this theory's argument that increasingly democratic societies have less inherent power structures that underline social relations, and that legitimacy is gained only through free and unconstrained public deliberation. She insists that power and legitimacy are entwined and that the concept of the political should be critically appropriated within the context of democracy. For Mouffe (2000b), political actors should take on the role of adversaries, who both share and simultaneously dispute a set of values as well as ethical and political principles. I will not explicitly be emphasizing power in theoretical terms, but it remains the underlying factor of contentious deliberation. The role of advocacy in creating a deliberative space is precisely what can diminish de facto power asymmetries among actors, enabling free and undistorted communication to flow (Habermas 1987, 1996). In this shifting face of representation, legitimacy has to be rooted in other avenues. In my case studies, I look at different advocacy groups within civil society and their attempts to derive legitimacy through mass participation. The initial onus of legitimacy is on government actors who worry about their legitimacy and seek it through involving a wider network to avoid the perception of capture by special interests (Lertzman et al. 1996, 127; Toke and Marsh 2003). In response, within this deliberative opening, advocacy groups represent the marginalized, far from the centers of power, and derive legitimacy from both the international and the domestic spheres.

The role of civil society in the deliberative system

Studies of deliberative democracy contain few critical discussions of the normative role of civil society in a discursive space. Hendriks (2006a) provides a review of two emerging ideas within deliberative democracy: the first is of micro deliberative theorists, who focus on the ideal conditions of a deliberative procedure (3) and do not directly refer to civil society. The second are macro deliberative theorists, who emphasize informal discursive forms of deliberation that take place in civil society (4). These two streams carve out very different roles for civil society. Micro theorists prescribe a role for civil society through participation in structured deliberative forums, in collaboration with the state, whereas macro theorists emphasize informal, unstructured political activity, beyond and against the state.

This division can lead to conflict between the ideals of a structured deliberative forum and the unstructured deliberations of civil society that seek to challenge the state and its institutions but often do not find a place within structured forums. Rather than attempting to merge these two distinct deliberative forms together, Jane Mansbridge (1999) suggests that we consider a "deliberative system." This system consists of a deliberative continuum, the extremes of which are differentiated by degrees of formality. At one end are discussions between citizens, social movements and interest groups that take place in public spaces (such as in the media) as well as in private spaces. At the other end of the deliberative system are formal decision-making institutions, such as public assemblies, ministries and parliament. Mansbridge acknowledges that her proposed system may be missing the "ideal deliberative procedure" set out by micro theorists, leaving out entire sections of civil society in the case of formal, institutional deliberative venues. However, the total system, she argues, achieves deliberative conditions (211). This continuum is a good starting point in assuring that civil society has a place in the deliberative politics framework. It also gives a fuller picture of how civil society engages in non-deliberative means to arrive at a formal deliberative structure as well as emphasize its roles in a structured framework.

Civil society actors play a crucial role in building public spheres and organizing members of the public, for several reasons. They challenge exclusions supported by the status quo, and they interrogate hidden assumptions that public policies may be rooted in and struggle for change (Fischer 2006; Polletta 1999). These public spaces are important as they allow civil society actors to frame and specify issues, flesh out meaning and ground their own experiences in ways that are less constrained by institutional forums. Advocacy plays an important role in building a deliberative continuum. I argue that even when a discursive arena is provided by the state, advocacy is still necessary in order for civil society to have policy influence. I outline some important pathways to effective civil society participation: the successful framing of issues, deriving legitimacy through international norms, and the formation of effective civil society alliances, through mobilization of grassroots actors. The case studies show two different forms of advocacy. The push for biodiversity policy came from the Convention on Biological Diversity (CBD), which provided a discursive opening for a small group of rights-oriented actors. The CBD shifted the axis of domestic rhetoric from conservation towards a people-centered, rights-oriented approach and made the National Biodiversity Strategy and Action Plan (2009) a unique top-down venture. The Scheduled Tribes and Other Traditional Forest Dwellers (Recognition of Forest Rights) Act, 2006, on the other hand, ensured rights and ownership of tribal and traditional forest-dwelling communities over common property natural resources. This was the culmination of a peoples' movement which some claimed had been fought over a period of 150 years, finally pushing on to the national stage the concerns of communities who had no policies protecting their rights.

Civil society organizations connect with governments and influence various stages of the policy process. Arnstein's (1969) classic ladder (table 5.1) shows the different ways people participate in the policy-making process and the reflected degrees of power and control. In between the bottom of the ladder, where the least power is concentrated, and the top, with the most power, there are intermediary rungs that include rubberstamp advisory committees, consultations in the form of public hearings, and partnerships where decision-making is shared. Boyle et al. (2001) differentiate between policy advocacy, which emerges from outside the boundaries of government, and policy participation, which occurs from within it. My research challenges this dichotomy by bringing advocacy that operates within and advocacy operating outside the government into the same frame. This dichotomy is collapsed within my research design where advocacy emerges within the walls of the government (within invited consultative processes) but is supplemented by strong advocacy outside the formal deliberative forum.

One of the important goals of deliberative governance is to give space to diverse voices of citizens who remain marginalized from the policy process (Phillips 1995). Thus, deliberative democracy attempts to move away from top-heavy centralized styles of decision-making by directly engaging citizens and experts in deliberative activities (Fung and Wright 2003; Weber 2003). In a legitimate democratic process, members of civil society can "push topics of general interest and act as advocates for neglected issues and under-represented groups," dialoguing with government and exerting influence on lawmaking (Habermas 1996, 368). Elster (1998) further emphasizes that the process of deliberative conversation does not simply involve discussing and arguing to convince another party of a particular view but also involves some sort of negotiation that involves the exchange of threats and promises.

Thus, while decisions continue to be made by the formal political process rather than by citizens, civil society nevertheless takes part in discussions that lead to decision-making. Ultimately, the mechanisms of participation and deliberation may help accomplish a key goal of critical theory – the emancipation of individuals and society itself from oppressive forces (Dryzek 2000). New ways of organizing, advocating and participating (e.g. internet campaigning, network engagement, affluence of new social movements) are part of the democratizing of contemporary political processes.

The deliberative democracy literature outlines three ways in which civil society organizations can communicate policy ideas to the state. First, they may cooperate directly with the state within deliberative forums or other cooperative arrangements (Montpetit et al. 2004). They argue that inclusive deliberation allows the state to incorporate "experiential" or "situated knowledge" (142). Second, civil society can exert influence indirectly on the state by the use of discourse and the creation of narratives in the public sphere (Dryzek 2000; Montpetit et al. 2004). Opinion leaders within civil society and the media can utilize critical discourse and connect rhetoric to specific government policies. In this way transmission of policy ideas occurs. Dryzek (2000) argues for the use of rhetoric as he sees its "ability to reach a particular audience by framing points in a language that will move the audience in question" (52). Its emotional appeal is "diffused and pervasive, felt in the way terms are defined and issues are framed" (101). In this role, civil society actors remain autonomous, confronting state actors from the outside, an autonomous space beyond the state. Third, civil society actors can combine these two positions in order to exert influence. Cohen and Arato (1992) elucidate a process where civil society organizations can both cooperate and confront state actors, in a "dualistic strategy."

I bring this "dualistic strategy" to the deliberative governance framework (see also Dryzek 2000; Habermas 1996) and argue that civil society organizations may enter deliberation in cooperation with the state as well as generate critical discourse to reframe policy outside the state's deliberative settings. This strategy is close to the idea of "forum shopping" (Thomas 2000) in which civil society actors take action outside the deliberative forum when dissatisfied with the process within the state perimeters. Few theoretical studies have considered the linkages between civil society and the state, and in particular how civil society attempts to transmit ideas to the policy-making space (Montpetit et al. 2005). This strategy forms a useful framework for explaining the tools of advocacy utilized within the deliberative governance space, though it is not restricted to specific forums. Here I map the "continuum" between micro and macro approaches, arguing that although policy is ultimately decided within the discrete deliberative space, the broader roles and activities of civil society actors do have bearing on policy processes.

Civil society and the state in India

The Indian state, since its independence in 1947, has instituted several legislations, policies and programs with the aim to improve the welfare of its citizenry, specifically targeting the more marginalized sections of society. However, the centralized nature of the state with its numerous functionaries undermined the very policies and programs it created between the 1950s and 1970s. The liberal-welfare roots of the state slowly eroded over the years, and its agenda was subverted to serve the affluent and privileged sections of the population. This led to greater alienation of the common people. It has been suggested that the colonial experience resulted in an emergence of a vibrant public sphere because of the role of popular movements and advocacy in the freedom struggle. After independence, however, the private spheres were captured by "native elites." This resulted in the cementing of identity-based politics, along the lines of caste and community, which came in the way of large democratization of Indian society (Ali 2001). As dominant groups and communities captured the benefits of development, the more marginalized sections of society who had suffered social and economic vulnerability took on all the costs of development and none of its benefits. The years following independence saw the phase of "nation-building" with investment being fueled into industrialization projects with an emphasis on large-scale irrigation, hydropower and heavy industries. The poorer and more disadvantaged sections of society bore the brunt of displacement. The state provided very few safety nets because of a persistent under-valuing of compensation for land and a lack of a comprehensive resettlement policy. This period saw the heavy commercialization of natural resources that were directed into the industrialization project. All the while, the sections and communities of people directly dependent on these resources for their subsistence lost access to their lands and continued to become more marginalized. In order to address the flagging public sphere, the Central Social Welfare Board (CSWB) was established in 1953 with the objective of promoting voluntary efforts in social welfare. This marked the beginning of government funding to voluntary organizations.

Civil society remained quiescent during this phase of nation building for almost a decade and a half after independence. There was faith in the nascent nation's role of provider, protector and regulator, and there was a consensus and expectation from the state to deliver. Towards the beginning of the 1960s it slowly became clear that the state had failed in its mandate to provide for all and was failing to provide the basic amenities of education, health, livelihood and shelter for the majority of its people. In addition, the state's failures were compounded by a severe crisis of food grains in the early 1960s.

Civil society initiatives in the 1960s found expression in many ways: some organizations created voluntary associations to fill the lacunae in state channels and provided basic services like health and education. Others organized activities with an aim to cooperate with the state and retained formal links with the ruling Congress Party. In the spirit of support and reciprocity, the government provided resources to these organizations in the form of offices, finance and structural support to strengthen their work (Tandon and Mohanty 2002). At the same time, students, workers and political movements also mushroomed in different parts of the country, and they posed a real challenge to the legitimacy of the state. The critical shift in the relationship between civil society and the state occurred with the imposition of the Emergency from 1975 to 1977, which saw the potential of a democratic state to turn dictatorial. This period saw severe curtailments in fundamental rights, the power of judiciary and the press:

> Emergency and the restoration of democracy not only redefined and extended the boundaries of civil society; it also by redefining the relationship of the citizens with the state restructured civil society in a significant way and made it more alert to transgression of its boundary by the state. The most important consequence of emergency for civil society was the question concerning the collapse of state institutions and their inability to protect the rights of the citizens.
>
> (Tandon and Mohanty 2002, 70)

The repressive political environment of the 1970s resulted in the emergence of collective mobilization in the form of widespread social movements. The experience of fighting colonial rule had equipped civil society to take on the state in the public sphere. These forms of collective mobilization centered on diverse issues and included the environment and feminist movements, the movement for *Dalits* (untouchables) and land movements around displacement of people as a result of large-scale developmental projects. These movements saw an emergence of new identities, stood in direct opposition to the state's infringement of fundamental rights and forced the state to make space for new concerns and voices. They challenged inequitable power relations and pushed for issues like ownership and control over resources, mainstreaming gender issues and the recognition of fundamental rights for traditionally powerless groups. Civil society activism in India is largely is rooted in the ideals of civil disobedience through non-violent means. It generally relies on mass rallies, petitions, marches and political agitation, all forms that were popularized in the struggle for Indian independence. These social movements differed from the older movements in terms of their "incisive analysis of those aspects of poverty and oppression – ecological degradation, subordination of women and so on – that are given short shrift by the class based movements" (Guha 1989, sec. 3.14).

The 1970s and 1980s also saw a new government coming into power (Janata Government 1977–80) who visualized special roles for voluntary organizations through its programs such as Adult Education, Integrated Rural Development Programme (IRDP) (Planning Commission 1978) and training programs of lower-level functionaries. Special exceptions and incentives were also granted to industries and businesses to involve voluntary organizations in their activities in rural areas. Encouraged by institutional incentives, the country witnessed the proliferation of professional organizations or NGOs and voluntary organizations (VOs). These provided services to the marginalized in the form of health services, micro-credit and rural banking services, primary education and sanitation (Mohanty and Singh 2001). To some extent, these were also conceived of as partners to the Indian state. For example the Seventh Five-Year Plan (Planning Commission 1985, 3.14; emphasis mine) states,

> Voluntary agencies have been traditionally working in the areas of relief and rehabilitation, education, health and social welfare. But they can also play a useful role in supplementing Government's efforts in other areas such as the provision of drinking water, release and rehabilitation of bonded labor, ground water surveys, development of alternative sources of energy and many other activities relating to rural development and poverty alleviation. Several voluntary agencies have acquired, over the years, professionalism and expertise to provide competent technical services and yet the services of voluntary agencies have not been fully exploited by governmental agencies for the implementation of programmes of welfare and poverty alleviation. This is partly because there is no institutional *forum where voluntary agencies and Government can come together. Such forums need to be established.* They will provide lines of communication between the official sector and the voluntary sector; also they will enable smaller village-based groups to receive funds from the Government and the Government, in its turn, would be able to obtain valuable information on the progress and problems of different development programmes.

In 1994, in a dialogue initiated by the Planning Commission, the government and voluntary organizations deliberated on the "Action Plan to Bring about a Collaborative Relationship between Voluntary Organizations and the Government," which was accepted as the basic policy that would govern the government and VO relationship in India. This idea of joint forums is particularly relevant to the process by which the Forest Rights Act and the National Biodiversity Strategy and Action Plan was negotiated.

The 1980s and 1990s saw rapid development in the voluntary sector because of a supportive government and growing funding by both national and international donors. In addition there was a changing conception of the development paradigm that centered on increasing peoples participation in the implementation of programs and management of resources. Although there is no complete survey or comprehensive study on the total number of VOs and NGOs working in India, some estimate that their number is about 100,000, of which only 25,000 to 30,000 are active. There are around 21,000 societies, which have been registered with the Ministry of Home Affairs, Government of India, under the Foreign Contribution (Regulation) Act (FCRA), 1976, since its inception in October 1999. Of the total registered societies under FCRA, more than 15,000 are such organizations that received foreign funds during 1998–99 (Mohanty and Singh 2001, 23).

Over the years, civil society has inhabited many different roles, but often it had been restricted to the implementation level of the policy cycle. Their role has generally been to ensure transparency, delivery of services, effectiveness and accountability of existing policies and resisting policies that are perceived as oppressive or impinging on fundamental freedoms. Policy formulation and agenda setting in India has traditionally been the purview of the national bureaucracy. More recent scholarship (Sapru 2010) suggests that CSOs can and do also play a role in policy formulation in several ways. Echoing Covey's five strategies – "education, persuasion, collaboration, litigation and confrontation" – I expand on some strategies that CSOs use to influence policy formulation:

Coordinating public opinion around a policy. Because civil society organizations serve as links between citizens and policymakers, they are able to coordinate large-scale campaigns to communicate about certain policies. In educating the public on the content and effects of certain policies and maintaining the intensity of public scrutiny and discussion on proposed laws, they are able to influence top-down policy-making from the bottom-up. "With access to information, civil society fosters democracy by limiting the state, providing space for protest groups, generating demands, monitoring excess, confronting power holders, and sustaining a balance of power between state and society" (Sapru 2010, 170).

Providing advisory services. CSOs often have expertise, be it technical or field experience in certain subjects or areas where they have established long-term engagement. By virtue of their experience, they are often able to provide technical data and evidence that feed into policies as well as weigh in on the feasibility of certain policies and their effects on local populations or landscapes. The government often responds positively to these insights and incorporates them into policy and sometimes even invites CSOs to contribute to the process of policy formulation.

Collaboration with policymakers. Some CSOs have the stature, credibility and power to directly influence the executive and its departments at the formulation stage, even before a bill is drafted. CSOs can be invited to draft policies and legitimately claim to represent an interest in the "common public good."

Taking the state to court. In the face of an unaccountable and unreliable state whose record on implementation of policies is weak, CSOs, who are unable to control or restrict the government's activities, often file public interest litigations (PILs). PILs are means by which civil society actors or individuals can sue the government and force agents of the state to respond to the queries and demands of civil society. Judicial activism is often lauded as the frontrunner in setting the shortcomings of the state right: "Social awareness, insistence on human rights and the attempt to check governmental lawlessness are said to have transformed the Supreme Court of India into a Supreme Court for Indians" (Baxi 1994, 143). CSOs see the PIL as an important social tool in bringing action on behalf of the community, advocating for new rights and changing exploitative laws and policies. On the other hand the inefficiency that haunts the Indian system extends to the courts as well, leading to the ambivalence in the perceptions of the court. They are seen "as simultaneously fountains of justice and cesspools of manipulation. Litigation is widely regarded as infested with dishonesty and corruption. But courts, especially High Courts . . . are among the most respected and trusted institutions" (Galanter 1984, 500).

Confrontation with the state. The deep pessimism that many (Lewis 1995; Kothari 1989) feel at the state of governance in India often leads to direct confrontation of the state with its people. This is manifested in CSOs becoming more assertive in confrontation with the state in the form of social movements that demands debate and transparency in policy-making in the public sphere. Social movements and collective action has played an important role in democratizing the state, articulating collective demands and pushing for more inclusive policies.

Inclusion. Faced with the centralized and authoritative power of the state, civil society plays an important role in bringing the concerns of marginalized populations to the public space. This space is a loose arena that is free and accessible to everybody. CSOs represent the interests of the marginalized, offer critiques of government and society, and empower communities to question and engage with the state. In addition, CSOs infuse a plurality of voices into the policy environment. This is done in many ways including the promotion of grassroots mobilization for social change (Clark 1991), participatory development (Bhatnagar and Williams 1992) and responses to public or private sector failures (de Tray 1990; Bratton 1988).

CSOs in India are a complex network of non-governmental organizations, self-help groups, voluntary associations and organizations, advocacy groups, religious associations, philanthropic groups and professional networks, amongst others. These organizations perform an array of functions from service delivery to serving as "watchdogs" that pressure the government into being more accountable to its people. Some organizations serve the interests of a particular community or group, while others take on crosscutting issues. Civil society in the context of India both has a collaborationist approach as it works with the government to tackle problems that are mutually agreed upon and at other times takes on an adversarial role and stands in direct opposition to the state. This background is important to keep in mind while looking at the case studies discussed in this book. One finds that the relationship between the civil society and the state is not homogenous, predictable or automatic but rather mutually reinforcing.

Strategies for creating political opportunities

The theoretical frames for the two policies represent two different advocacy strategies of NGOs and CSOs in the face of state-led policy processes. The two strategies are rooted in different impulses – norm diffusion is an exogenous process where advocacy is organized around ideas that are transmitted from beyond the limits of the state. These are then translated and deliberated within domestic forums in an effort to create pressure on the state to open up policy to different interpretations. Mobilization, on the other hand, is an endogenous process where pressure is created by people on the peripheries of the state organizing and pushing its agenda into the deliberative space of the state. The final word on both policies is with the state; however, civil society and NGO advocacy succeeds in bringing different viewpoints and interpretations to the table providing fuller and more equal terms for deliberation. This model is illustrated in figure 2.1.

Civil society, in its multitude of definitions, has become a key in the language of governance. It has become an important part of "delivering services, substituting democracy and representing a crucial international link" (Esteves et al. 2009). It has been seen that many international agreements concerning the provision of financial resources, the financing of development projects and the liberalization of trade require guarantees of good governance, including the participation of stakeholders in the policy-making process (e.g. Monterey Consensus; infra). Social movements represent acting outside the consensus, pushing through agendas by acts of protest or action. Civil society plays a role in deliberative democracy, not only by articulating their concerns but by finding strategies to convey their experiences of social problems to the broader public and to the state. These strategies can be understood first as linking specific interests of particular groups to international norms of rights, which are then deliberated within state-sanctioned forums, and second as creating social movements through large groups, highlighting the rights and issues of particular groups and bringing them to a national space where the conflicts with more general interests can be deliberated on.

Scholars of discursive democracy emphasize their contributions to a vibrant public sphere in which discourses are disseminated and refined (Torgerson 1999). Not only does civil society emerge as a site for discourse or "new discursive horizons" (Dryzek 2000, 140), but civil society actors also play a role in directly challenging state policies that are viewed as illegitimate (Stivers 2002; Young 2001). To this end, two strategies are outlined that bring issues to the few deliberative spaces provided within the state

Changing the conversation from beyond the state

Constructivism makes claims to the nature of social life and change, pointing to ways to make the link between structure and agency. At the heart of this are

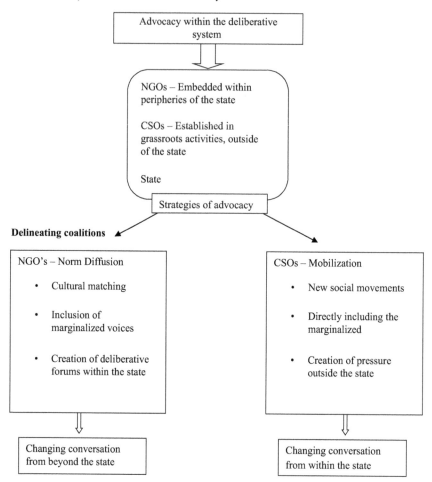

Figure 2.1 Conception of the theoretical model

agents who translate ideas into normative structures. Constructivists are especially interested in how political actors produce the inter-subjective understandings that undergird norms (for example, see Risse et al. 1999; Barnett 1999). This pre-occupation with interpretation can also be related to the "argumentative turn" within deliberative policy analysis, which draws on insights of constructionist epistemologies and argues that the policy-making process is a political contest over the meanings of policy problems and concepts rooted in the arguments that people make. Though deliberative mechanisms are still considered legitimate forms of governance (Hajer 1995), negotiation over political and social meaning are still the essential focus of policy analysis. Thus, public contestation and construction of meaning is still the focal point in using both deliberative forms and discourses (Healey 1993). Agents can remake structures, spurring political change by contesting meanings, inducing shifts in identity and creating and demanding the space for change. Norm entrepreneurs are these very agents who have been defined as "specialists who campaign to change particular norms" (Hechter and Karl-Dieter 2001, 45).

Attention has been given to communication, especially persuasion, which attempts to change actor preferences and to challenge current or create new collective meaning. Individuals and collectives closely relate this communication to Dewey's view of public deliberation in which democracy is a method for overcoming problems faced. Democracy continues to be imbued with the type of discussion that requires participants to consider their actions, which are based on beliefs, and their relational position to the views of others (Dewey 1984; 1986). Political scientists argue that meanings of particular norms and the linkages between existing and emergent norms are often not obvious and not "persuasive connections" and have to be actively constructed. All advocates of normative change confront "highly contested" contexts in which their ideas "must compete with other norms and perceptions of interest" (Finnemore and Sikkink 1998, 897). This also fits in with the communicative aspect of advocacy by which persuasive policy narratives are constructed, disseminated and debated.

Different waves of scholarship have focused on distinct areas of domestic-international norm linkages. The first wave had little analysis about the contention between norms and counter norms, and between and among local, domestic, and international actors. This perspective ignored "ingrained beliefs that are inconsistent with the persuader's message" (Chekel 2001, 222). The "second wave" of norms scholarship tried to analyze domestic politics more directly, using domestic structures to emphasize political conflict and institutions and culture as "filters" through which international norms are transferred. These filters are also related to framing exercises. International norms are more likely to be used if their meanings are framed in a way that they can be interpreted to be different things to different actors. It is clear that norms at an international level should not be studied as monolithic. They exist in different variations, nuances and even conflicting principles that are not consistently adopted by the state in question. States have a possibility to pick and choose the norms most appropriate for them: those that may be consistent with local values, practices and beliefs, for instance. Norms that emerge out of very specific local conditions that are directly adapted to different polities – or norms modeled on different political cultures that cannot be translated – are often unable to gain influence (Subotic 2009). Martha Finnemore (1996) contends that states are socialized to want certain things by international society. Framing these demands within the constructivist approach, she develops a systemic approach to understanding states' changing behaviors and interests by investigating the international structures of meaning and social value. State interests, she argues, are defined by internationally held norms. This normative context influences the behavior of decision makers as well as mass public who may choose to constrain them.

Scholars suggest that the successful diffusion of a particular international norm requires a degree of congruence between the norm and the domestic conditions of states. If there is a lack of congruence between international norms and domestic structures, domestic actors may attempt "congruence-building" by localizing a norm or rejecting it outright (Acharya 2004). Two processes through which congruence may be established are "framing" and "grafting." Payne (2001, 39) claims that a frame is a device used to "fix meanings, organize experience, alert others that their interests and possibly their identities are at stake, and propose solutions to ongoing problems." Framing an issue within a certain cognitive schema is an essential component of advocacy as new frames resonate with wider public understandings and are adopted as ways of talking about or understanding issues (Finnemore and Sikkink 1998, 897). "Grafting," meanwhile, is defined as "institutionaliz(ing) a new norm by associating it with a preexisting norm in the same issue area, which makes a similar prohibition or injunction" (Acharya 2004, 244). Congruence building is an iterative process of the norm-building process as both conditions and ideational structures of the domestic or international levels evolve.

A broader literature review reveals two different mechanisms at work in the process of norm diffusion: the "bottom-up" and the "top-down" processes. The former has two paths: first, domestic social actors, even in isolation from broader transnational ties, may exploit international norms to generate pressure on state decision makers (Cortell and Davis 1996). The latter illustrates how non-state actors come together with policy networks, at both the national and the transnational level, united in their support for norms; they then mobilize and coerce decision makers to change state policy (Keck and Sikkink 1998; Subotic 2009; Charney 1993; Brysk 1993; among others).

Subotic (2009), surveying the top-down diffusion mechanism, points to social learning rather that political pressure leading agents (identified as elite decision makers) to adopt prescriptive norms, norms which become internalized to the extent of being behavioral claims. This process is based on notions of complex learning, rooted in cognitive and social psychology, in which new interests are taken on by individuals, when exposed to the prescriptions embodied in norms (Haas 1990; Finnemore 1996; Checkel 1997). In summary, norms can play two important roles in domestic politics. At one level, civil society can exploit spaces created by international debates to pressure domestic elites and reframe debates, garbing old issues in new terms. However, norms may also help agents expand their cognitive visions, by giving them the opportunity to adapt to new preferences (Checkel 1999).

Many international normative interventions are aimed at societies where there is a limited demand for normative change. International actors continue to push for certain institutional solutions because they believe that some norms cannot be violated and policies promoting them are unacceptable and because they agree that domestic policy change is necessary to accommodate new norms and that societies and their elites will eventually be socialized or eased into accepting international rules over time. Domestic demand from citizenry for normative changes, however, depends significantly on the character of the international intervention. For instance, if there is a lack of congruence between the proposed normative shifts and broadly shared values at the domestic level, international intervention will be resisted while the demand from below for change will remain low. Thus, in the discussion of values and shared consensus, domestic culture has an important role to play in the acceptance of norms.

A cultural match as defined by Checkel (1999, 87) is "[a] situation where the prescriptions embodied in an international norm are convergent with domestic norms, as reflected in discourse, the legal system (constitutions, judicial codes, law) and bureaucratic agencies (organizational ethos and administrative procedures)." In the case of a cultural match, domestic actors are likely to put more weight on international norms, recognizing the norm's associated obligations. Conversely, the salience of foreign norms in domestic policy can be hindered by the lack of a cultural match (Checkel 1999). If the international norm conflicts with the domestic culture, domestic actors may resist international pressure to adopt that norm. The state may find the acceptance of the international norm as compromising its sovereignty (Cortell and Davis 2000). This match is not just about traditions and customs but also extends to broader ideologies – perceptions shaped by the socio-cultural environment as well as individual belief systems. In studying advocacy in the context of norms, this culture plays an important role in determining the strength and activism of their demands.

This context is crucial in understanding the particularities of the case studies. The macro-level National Action Plan and Strategy on Biodiversity was envisaged in 1994 when India began the process of drafting the biodiversity bill. This happened when a core group of individuals, including NGOs and research institutes, evaluated the existing legal framework and recommended the development of a law dedicated to the objectives of the Convention on Biological Diversity (CBD) (Anuradha et al. 2001, 2). The preparation of the NBSAP was a mandatory requirement under the UN Convention on Biological Diversity, of which India is a signatory. The process was initiated in 1999, when the Ministry of Environment and Forests (MoEF) of the Government of India was granted funds from the Global Environment Facility (GEF), through the UNDP (UN Development Programme), and the formulation process was formally launched in January 2000. This process has looked at biodiversity in all its forms: natural and agricultural ecosystems, species of wild plants and animals, microorganisms, crops and livestock, and their genetic diversity. Aspects of conservation, sustainable use of biological resources, and issues of economic and social equity were also covered. The NBSAP process involved the preparation of seventy-one Biodiversity Strategy and Action Plans (BSAPs) at the local, state, eco-regional and thematic levels, and also thirty-two sub-thematic review papers. The preparation used the norms of the CBD and the financial support of the UNDP to create a process that was highly participatory in nature, reaching out to village-level organizations, CSOs, academicians and scientists, government officers from various line agencies, the private sector, the armed forces, politicians, artists, media persons and others. Over 50,000 people were involved in various capacities (Kothari 2005).

Applying theoretical understandings about domestic use of international norms is important because it explains why it is much easier for a state to go through the motions of complying with international norms and simply ratifying institutions than to make them work effectively in a domestic context. It is usually much easier for domestic actors to symbolically comply with international norms than to bring about profound social change that these norms require. Thus, it is important to identify how domestic use of international norms is to be predicted and incorporated into our understanding of how international norms become diffused and to what domestic political effect (Subotic 2009).

To understand how norms diffuse or how specific coalitions of actors who aim to change policy succeed or fail, it is important to study domestic political structures of different states. Risse-Kappen explains a subjective view of "state autonomy," which depends on different contexts and degrees of the state's distribution of power and its embeddedness in international regimes and organizations. This explanation is derived from domestic structure approaches that allow differentiation between various degrees of state strength and autonomy vis-à-vis society and does not limit it to merely "statist" or "pluralist" approaches (1995, 18). Domestic structures refer to the political institutions of the state, to societal structures and to the policy networks linking the two. Domestic structures encompass "the organizational apparatus of political and societal institutions, their routines, the decision-making rules and procedures incorporated in law and custom, as well as the values and norms embedded in the political culture" (Risse-Kappen 1995, 19) and is explained through the axis of state structures (centralized versus fragmented), societal structures (weak versus strong) and policy networks (consensual versus polarized).

India has been identified as having the characteristic of a strong (centralized) state facing strong social organizations, which often leads to "stalemate" situations, exacerbated by a highly polarized polity and a political culture emphasizing distributional bargaining. Access of transnational actors might be easier than in a state-dominated case. However, the policy impact could be limited due to structural problems of societal and political institutions that may deter policy change. It is expected that change that occurs will necessarily arise from domestic processes that alter domestic structures during the process (Risse-Kappen 1995, 26).

Cohen and Arato (1992) observe that the participation of civil society associations in the public sphere is the defining trait of contemporary social movements. They contend that traditional modes of organizing into social movements, based on mass action and collective behavior, are largely outdated and bring out irrational aspects of human behavior. In contrast, the rationality of deliberative democracy encompasses new kinds of collective action and movements in the context of pluralistic civil society. Hendriks (2006a) offers a conceptualization of the deliberative system "where communicative practices that foster critical, public reflection take place" (499). Discursive spheres "are sites where public discussion occurs through the exposition and discussion of different viewpoints" (499). These include forums, arenas, courts and the public sphere. These spheres are "embedded in a broader, informal 'macro' discursive context that includes the competing discourses available to any actor" (499). The "macro discursive context" is the sphere of civil society or media where the contestation of discourses takes place.

Civil society organizations may participate in structured forums to transmit policy ideas based on their local knowledge (Monpetit et al. 2004) or disseminate critical policy ideas through discourse in the public sphere (Dryzek 2000; Warren 2001; Cohen and Arrato 1992) or extra-constitutional spaces in national or international spheres (Dryzek 2000). I focus rather on these autonomous spaces, as overlooking them undermines sources of democratic vitality and equality (Fraser 1990). These are the spheres within which deliberation occurs. The macro-level action plan acted more as a policy guide, reflecting which direction the law should take rather than being legally binding. The conceptual frame of the plan was left to the Technical Policy Core Group (TPCG), which included researchers, activists and representatives of NGOs who were invited by the government to collaborate on collating a participatory biodiversity strategy. Its use of mass participation, engagement with grassroots interpretations of biodiversity and norms enshrined by the CBD, inclusion of values regarding biodiversity, and use of the deliberative space to launch a critique of top-down governance strategies made it interesting to study. This case made it evident that there is a shift of dynamics within "invited" spaces of participation as well as important institutional and structural responses to these shifts. Though many scholars have given neat divisions to the tensions between social movements and civil society consultation, the realities are far more blurred. In this book, we compare consultative mobilization that is incorporated by the state but attempts to act outside it with social movements that become co-opted by the state and turn consultative.

Hendriks (2006b) associates civil society actors (which she sees as interest groups, social movements and the like) with the macro discursive sphere and government actors with the "micro discursive context." This is a viable conceptualization, though one can point out that government agents and civil society actors can participate in both the macro and the micro discursive spheres. The interaction of these spheres is often overlooked in the theory of deliberative democracy, but it is crucial for understanding how policy ideas are advocated to the state by civil society. The deliberative forum is only one part of this system, especially in the process of policy-making. This forum is embedded in a much larger context that shapes what happens within a deliberative structure. Thus, the context remains as important as the deliberative opening.

Changing the conversation from within the state

Governance has incorporated new structures of democratic practice that complement conventional modes of public participation with the creation of new spaces, laws and policies. In emerging countries like India where democratic practice has had time to take root, the thrust for more public voice within policies comes in different and sometimes conflicting spaces. Especially in policies governing resource use, the underlying emphasis seems to be that voice must be given to the resource users. Linking the notion of deliberative politics to a public space, Hummel (2002) notes that public space needs to pose questions about the underlying assumptions people hold about what society is and what it is doing. "Democracy" as a way of organizing the state is understood as "territorially based competitive elections of political leadership for legislative and executive offices" (Fung and Wright 2003, 3). Yet, mechanisms of political representation seem increasingly ineffective in fulfilling central ideas of democratic ideals: full and equal participation of the citizenry, transparency, dialogue, the implementation of sustainable public policies that promote equality and an equitable distribution of a nation's wealth.

The modern state has been defined as "an amorphous complex of agencies with ill-defined boundaries, performing a great variety of not very distinct functions" (Schmitter 1985, 33). Two waves of research in American political science have both confirmed and rejected this view. The first abandoned the state with its own conceptual identity and replaced this with the idea of a political system. However, this change in semantics could not erase the conceptual idea of the state as distinct from the other social and cultural systems it interacted with. Since the 1970s, there had been a movement to "bring the state back in" (Evans et al. 1985). This wave of scholarship set about separating the state from society and giving it a partially or wholly autonomous identity.

Linking the idea of the state or parts of this political system encouraging social movements also brings us to the idea of consultation and the role of the state. If we see the state with separate dynamics creating space within it for consultation with civil society, one must mention that, historically, popular self-organization has in general been discouraged by the ruling elites. This plays an important role in determining how democratic institutions encouraging consultation are different from social movements, where self-organization is outside the shadowy limits of the state.

Many social movements that are concerned with creating more participatory space are repositioning citizens within more "traditional" forms of government and positioning them as an individual or collective voice that needs to be heard by the modern state. In movements, this is done in a larger scale and is often carried by its own self-generating momentum. The role of CSOs is visibly important as well, as it creates interfaces with a diversity of institutions: state, international donors, and corporations for leveraging in spaces that were once closed off to citizens. Participation thus comes to reconfigure relationships and responsibilities in an expanded public arena with a host of other actors (Edwards and Gaventa 2001; Tandon 2002). Cornwall (2002) reflects on the spatial dimension of participation, as an act of creating space for more opinions, people and deliberation, bringing spaces to life as well as creating new social forms with their own momentum. Others (Lefebre 1991) see the public arena in modern democracy both in an abstract idea of "space" and also more concretely, in terms of actual sites used by citizens.

Equating the notion of democracy with full and equal participation brings us to "participatory governance." This idea is rooted in the notion that public policies have to be discussed by a variety of stakeholders who all represent specific interests. These stakeholders include a variety of different organizations, including trade unions, social movements, grassroots organizations and NGOs, which taken together represent a diverse range of voices and issues. Civil society is not homogeneous. NGOs and social movements are two concepts that are both part of a wider notion of civil society but have clear distinctions. However, NGOs often form part of or lead social movements. NGOs are value-based and increasingly professionalized organizations, which depend on donations and voluntary service rather than regular contributions from their members. Social movements are comprised of informal affiliations, groups, individuals and organizations united by a set of common political goals or social beliefs, whose aim is to bring about social change and influence the agenda of formal or informal institutions. They voice a "series of demands or challenges to power holders in the name of a social category that lacks an established political position" (Tilly 1985, 735) or adequate representation.

Mobilization has been widely recognized as a key option for political participation by the mainstream research on citizen politics (Rosenstone and Hansen 1993; Verba et al. 1995) and as one of the central functions for social movements and other political intermediation organizations that promote collective action (McCarthy and Zald 1977; Snow and Benford 1988). In classic studies of social movements, there has been a distinction made between instrumental and expressive movements (Searles and Williams 1962). Instrumental movements are aimed at a particular external goal, while expressive movements are a goal in themselves, voicing of protest against injustice. Protest mobilization often exceeds the narrower definition of campaigning for specific interests and political goals and includes opportunities for non-instrumental participation.

The distinction between instrumental and expressive dimensions in social movements brings in broad political repertoires and a corresponding expansion of political spaces from narrow electoral dynamics into alternative action opportunities played out in multiple arenas (Habermas 1981; Giddens 1991). Non-electoral participation arenas may engage citizens, who were previously uninvolved, into more direct political participation. As non-electoral participation slowly becomes appropriated by conventional modes of political deliberation, the contentious nature of protest is reconfigured within the acceptable boundaries of political dialogue. In this scenario, mobilization does not merely support collective action but also performs the complementary function of political communication, which legitimizes the process (Habermas 1994). This communicative feature within protest mobilization is particularly significant as it encourages a greater compromise than conventional participation. However, expressive protest also needs to be connected to more general or mainstream interests when causes are publicly exposed. This communicative element is also apparent in McAdam et al. (1988, 709), where there are two functions to the mobilization process: providing a context for collective action through the framing of particular issues and providing a staging ground where individuals meet through communication networks. These functions both rely on public dialogue as a form of connecting individual interests to collective concerns and serve as vehicles into public or formal spaces where deliberation occurs. These staging arenas provide a space in which previously independent spheres may coincide. These include the discourse of the state represented by ministries, decision makers, technocrats, policy experts and civil society.

Conover (2003) argues that if group identities are linked to particular identities, they can undermine public reason by focusing narrowly on specific interests in contrast to general interest. This is both proved and over-turned by the two case studies in India. Though biodiversity policy focused on the rights of livelihood of a particular group of resource users, it did attempt to link the group's narrower self-interest to larger public concerns on biodiversity preservation and state concerns of bio-piracy. The advocates of the Forest Rights Act, however, did not seek to present their collective action as a general public concern except to link it to a general universal idea of human rights. They argued that the specific interests of the forest-dwelling tribal communities should take precedence over general public ideas where the state controlled all conservation. The contention arose when adversarial groups argued that preservation of natural resources would best be done following a model where the resources could be protected better and with more of an overview if the state retained control of these resources. This created layers of tension, between group identities and rights and national priorities, as well as specific group rights in conflict with universal rights. For the goals of civil society, both in a consultative role and in a more antagonistic role, creating social movements is a viable strategy. Both the consultative role and social movements have strategies, agendas and spheres of action that occasionally overlap. Civil society's interest in advocating for participatory governance is to link marginalized voices to a space that they have remained excluded from. NGOs who act in a consultative role bridge these two spheres of activity and act as a liaison between the language of the people and the language of the state. This takes place within the state-given institutions and therefore within its framework; NGOs can negotiate access and inclusion within the clear parameters of citizen rights and laws. Social movements, however, resist patterns of development that have led to people's impoverishment, while NGOs seek to reinvent the very laws and strategies that the state is armed to negotiate with. The movements deepen democratic practices by replacing the exclusionary narratives of development with a counter-narrative of inclusion; expanding non-party political space for social action; democratizing the public space by saving it from being hijacked by the state; renegotiating people's relationship with the state; and bringing the discourse of citizenship and rights to development discourse (Mohanty and Singh 2001).

With the advent of liberalization in the 1990s, the Indian state repositioned itself with regard to civil society, blurring the lines of exclusion and extending collaborative partnership with representatives of civil society. More and more civic organizations were invited to assist in implementing state-driven social development programs. At the same time, the state opposed any social movements that struggled against any state-led activities in the creation of special economic zones, mining or dams. This created fractures not just between the state and civil society but also within civil society itself, which struggled with loyalties of collaboration with the state and voicing protest against it.

After independence in 1947, the Indian state tried to marry the agenda of development with that of democracy, one that would form the foundation of a society based on values of egalitarianism and social justice. A socialist element was built into the democratic agenda which development was to fulfill (Planning Commission 1951). Through the initial decades of the five-year plans, economic growth through industrialization and commodity production was envisioned as the core of the Indian economy. The face of economic growth was in large-scale projects of irrigation, mining and industry that began to transform the social landscape. For almost two decades, industrialization was accepted as a strategy of national growth. These initial decades of acceptance gave way to dissenting voices from the 1970s onward. Resistance movements focused on three main problems: (1) the models of development set up by the Indian state benefitted only a section of the people; (2) only elites could negotiate with the state for resources, and communities dependent on natural resources for their livelihoods were not given any alternatives when they were displaced or expelled from the resources in the name of a collective "development"; and (3) development was heavily centralized with the state making no space for negotiating or discussing policy with civil society. Industrialization was conceived of by the state as "public good," for which the state appropriated people's resources. One example of this was the taking over of people's land under the Land Acquisition Act of 1894. Thus, the state decided on the priorities of economic growth over people's resources (Mohanty 2007). Even now, state response to civil society activism remains "selective." The state is protective about its framework, strategy and goal. Only when civil society's framework, strategy and goal align with that of the state, or at least do not differ radically from it, is the state is willing to acknowledge them as legitimate.

The comparisons between civil society and social movement have come from two related impetuses in India. The growth of the NGO sector and the rebirth of social movements in India began in the 1980s. As NGOs have become institutionalized, some scholars argue that some NGOs are now almost indirect parts of the state that have lost their critical distance to the state because their funding sources are the same institutions they once challenged. This has been accompanied by a shift in the state system towards the increasingly globally accepted idea of liberal democracy. Although in many cases this is purely a formalistic move, the state that until the 1980s resisted the inclusion of the popular classes has been forced to concede far greater space to popular inclusion. In India, the 1980s saw the birth of a host of voluntary organizations that took on the entrepreneurial role of consultation with the state, addressing issues of ecology, rural development and health. These organizations took up roles at all levels of the state; the legitimization of grassroots knowledge in informing political decision-making also began at this juncture (Pant 2010). Many scholars (Esteves et al. 2009) have pointed out that this has led to a "domestication of popular inclusion" and an increasing tension between the prolific versions of participation in the form of "consultation" with NGOs and civil society taking participation into their own hands for their own purposes and in their own forms. This is further illustrated in the two cases.

The Scheduled Tribes and Other Traditional Forest Dwellers (Recognition of Forest Rights) Act, 2006, was passed in India on December 18, 2006. It has also been called the "Forest Rights Act," the "Tribal Rights Act," the "Tribal Bill" and the "Tribal Land Act." The law concerns the rights of forest-dwelling communities to land and other resources. The bill came about because of a 2002 government order of eviction of forest "encroachers." Activists who had been fighting at the grassroots level mobilized around these eviction drives to push for a legislation of rights and claims of forest-dwelling and tribal peoples to their ancestral land, which the government had recorded as "forests" and often included large areas of land that were not and never were forests at all. The process included mass mobilizations at the grassroots level with representatives of the movement entering deliberation with the government, being part of the Technical Resource Group (TSG) and aiding in formulating a law to achieve justice and empowerment for marginalized forest-dependent communities. Several conflicts tied to the idea of forests and their management plagued the formulation of the act. The legislative process became a site of conflict where different viewpoints and actors with deeply entrenched ideas on conservation, management and rights to the forest met and confronted each other. Tribal rights activists termed the passing of the bill a "watershed event" because through it the forest-dwelling communities would get administrative rights in forest management for the first time in the history of Indian forests (Ghosh 2006); environmental bureaucrats and other conservationists, on the other hand, termed it as an "ideal recipe" to ensure the destruction of India's forests and wildlife by "legalizing encroachments" (Krishnan 2007).

In the case of the Forest Rights Act, forest-dweller activists pushed for the rights of Adivasis (India's indigenous tribal population) to have their rights over traditional land reconstituted. It was initiated in 2006, and, after many debates and discussions in different forums, the bill was notified into force in 2008. The task of drafting the legislation was assigned to the Ministry of Tribal Affairs (MoTA), which constituted a Technical Resource Group, consisting of representatives of various ministries, civil society and legal specialists to draft the Scheduled Tribes (Recognition of Forest Rights) Bill, 2005. Guided by facilitators, advocacy at crucial moments and venues of the policy process, adapting advocacy strategies to different forums, as well as large-scale mobilization of stakeholders allowed for the passage of the bill into an act that came into force in 2008.

Pressure mounted on the government by tribal bodies and key political allies to introduce structural changes in favor of the forest-dependent people and resulted in the opening up of the policy process to deliberative structures, for collaborative policy formulation. Like the NBSAP, the conceptualization of the act included researchers, activists, social movements, individuals and non-governmental organizations, which were invited by the government to collaborate on and discuss the Forest Rights Act in formal deliberative spaces.

Thus, we find that although the two impetuses for the case studies were different, civil society played comparable roles. They played an important role in linking particular group rights to both general and universal norms. The pathways for the idea around which these identities were linked ultimately also determined their strategies. It is on these strategies that the case studies will shed light. What we see in the previous section is the theoretical frame around which these strategies work and the reiteration of an expanding space that civil society begins to occupy vis-à-vis the state.

3 Deliberating on the National Biodiversity Strategy and Action Plan

Can global norms form an effective basis for advocacy at the domestic level? I argue that India's ratification of the Convention on Biological Diversity (CBD) strengthened the leverage of rights-based domestic actors and allowed for diverse actors to be included in the formulation of India's biodiversity policy. I analyze different coalitions that advocated for marginalized rights, and show how they influenced the formation of the National Biodiversity Strategy and Action Plan. This case study makes an important contribution to the influence of global norms on domestic deliberative politics because it refers to a debate on an issue which was not salient in Indian politics before the ratification of the CBD. Despite the relevance of issues surrounding biodiversity, the specific agendas of sustainable use and equitable sharing of the benefits were not given much consideration in policy in India. Efforts to conserve India's biodiversity have evolved over the decades, but under the concept of "conservation" and not biodiversity preservation per se. Although 5 percent of the country's surface area is legally protected, the entire premise on which resource governance is based has attracted a considerable amount of criticism because of the exclusion of people from planning and conservation strategies.

Since the mere ratification of conventions does not lead to change, much of this influence is normative. A significant part of the literature on global norms identifies local salience as a crucial variable; global norms work when domestic and public institutions are open to their influence (Cortell and Davis 2000). Analysis in the biodiversity policy space in India shows that the norms emanating from the CBD were broad enough to be absorbed by the pockets of advocacy within the domestic space that kept the debates on rights alive. For the first time, a rights-based discourse combined with public deliberation gained currency in the Indian policy arena. Global rights norms and treaties can provide and create a more conducive environment for bringing actors together for advocacy and policy-making in areas that were not previously considered priorities by the state. This reflects the rights "turn" in international relations and a growing awareness that ideas and principles combined with material interests play a role in the dynamics of global politics (Guzzini 2000; Adler 2000; Cortell and Davis 2000; Checkel 2001). The growing viewpoint that ideas are significant in their own right also rests on an increasing awareness of the importance of a principled civil society organizations, motivated not only by materialism but by moral concerns in political life.

Biodiversity policy formulation in India took a form that was distinct from the top-down, hierarchical arrangements of formulation that characterizes the Indian policy arena.[1] During the policy-making process, the concepts of participation, sustainable use and equitable sharing of benefits became ingrained in the understanding of what biodiversity policy is. In addition, the National Biodiversity Strategy and Action Plan (NBSAP) followed an unusual policy formulation process with a definite emphasis on more decentralized processes and inclusive participation, which conformed to the mandates of the Convention on Biological Diversity. India had never framed any consistent biodiversity policy prior to this. Instead the question fell under a variety of different jurisdictions: forest law, trade agreements, patent policies, wildlife acts and agricultural policy. The thrust to push for an integrative biodiversity plan came from the evolution of the debate in the international arena. Much like the United Nations Conference on the Human Environment of 1972, which forced India to put environment on the policy agenda, the CBD initiated a similar process of reevaluation with biodiversity.

The CBD was crucial in fostering deliberation in the policy space in India. The CBD, combined with concerned domestic non-governmental organizations such as Kalpavriksh, highlighted people's rights politics in India and reshaped the terrain of domestic advocacy. First, the CBD promoted a rights-based discourse and delegitimized alternative approaches to issues surrounding biodiversity in India. It provided a discursive opening for a small group of rights-oriented actors, shifting the axis of domestic rhetoric from conservation towards a people-centered, rights-oriented approach. Second, it encouraged the formation of a rights-based civil society coalition that, once consolidated, turned the CBD from a mere declaration of principles into the inspiration for a rights-oriented inclusive politics aimed at introducing legislative and institutional reforms. However, the process was also a political failure with the state turning against the biodiversity plan in the last stages of the project. Thus, we will see that the narrative of rights-oriented inclusive politics can also be an obstacle for deliberation, given that global discourses also strengthen the centralized "management" of resources for the state. This case study studies the success of norms in being absorbed by pockets of advocacy and the opposition it meets by the state.

Background

Biodiversity-oriented processes in India

In the period before the CBD, India never framed a consistent biodiversity policy. Instead the issues related to biodiversity fell under the purview of different laws. Initially, the Indian Wildlife Protection Act of 1972 protected the majority of India's biodiversity. It was amended in 1991 (Wild Life [Protection] Amendment Act, 1991) and then again in 2002 (Wild Life [Protection] Amendment Act, 2002), but it still did not cover the entire gamut of genetic resources.

Some of the important milestones in biodiversity conservation in India were the policies in Table 3.1. Others like the 1988 National Forest Policy had conservation as a fundamental principle in the policy: "Conservation includes preservation, maintenance, sustainable utilisation, restoration, and enhancement of the natural environment" (MoEF 1988, preamble). These echo the core norms of the Convention on Biological Diversity. By 2009, the government had set aside 4.8 percent of the geographical area of the country for the exclusive conservation of its biodiversity in the form of protected areas. The 659 protected areas included 100 national parks, 514 sanctuaries, 41 conservation reserves and 4 community reserves (MoEF 2009, 19). In addition it enacted the Environment (Protection) Act, 1986, and Foreign Trade (Development and Regulation) Act, 1992, for regulation and control of biodiversity.

Table 3.1 Pre-CBD policies (adapted from MoEF 1998, 4)

Indian Forest Act, 1927: enables the state to acquire ownership over forests and their produce and, specifically, to facilitate trade and timber. The main focus is on controlling and regulating the timber trade.

Wildlife (Protection) Act, 1972, amended in 1983, 1986 and 1991, 2003: provides for the protection of wild plants and animals and regulates hunting, trade and collection of specific forest products. Certain tribes are, however, allowed to pick, collect or possess specified plants for their personal use. The revised act also provides a licensing system to regulate cultivation and trade of specified plants in a pattern similar to the trade in fauna. Licensees are required to declare their stocks and follow prescribed procedures.

National Wildlife Action Plan, 1973: identified broad goals of establishing a network of representative protected areas and developing appropriate management systems which take into account the needs of local peoples and conservation requirements outside protected areas.

National Forest Policy 1988, as amended in 1989: stressed the sustainable use of forests and the need for greater attention to ecologically fragile, but biologically rich, mountain and island ecosystems. It encouraged native tree species, customary rights and concessions for tribal communities.

Forest (Conservation) Act, 1980, amended in 1988: primarily deals with using forest lands for non-forestry purposes, mainly industry and mining. It requires state governments to acquire the approval of the central government before it de-regularizes a reserved forest, leases forest land to a private person or corporation, or clears it for the purpose of reforestation. Implementation of this act has reduced the annual rate of diversion of forest lands for non-forestry purposes to 16,000 hectares a year, compared with 150,000 hectares per year prior to 1980.

Environment (Protection) Act, 1986: empowers the central government to take appropriate measures for the purpose of protecting and improving the environment. It is authorized to lay down environmental standards for controlling different aspects. In accordance with this act, the central government has issued a number of regulations affecting sectors such as hazardous and chemical wastes, genetically engineered microorganisms, and industrial development of coastal zones.

Foreign Trade (Development and Regulation) Act, 1992: designed to stimulate sustained economic growth and enhance the technological strength and efficiency of Indian agriculture, industry and services. Import and export are prohibited/ restricted through licensing or routed through specified agencies.

India has a wealth of undocumented wealth relating to the conservation and use of biodiversity. In 1982, the Department of Environment initiated a project All India Coordinated Research Project on Ethnobiology (AICRPE) to identify and document indigenous knowledge. Its broad objective was to preserve the knowledge system of tribal communities. While there were considerable legal inputs present for conservation of biological diversity "in situ," with respect to creation of national parks or trade in endangered species of flora and fauna, there were widespread weaknesses in the legal frameworks with respect to conservation of domesticated or agricultural biodiversity. Even in the wake of being a signatory to the CBD, it was found that many ideas central to the CBD had no legal support structure in India. These included equitable sharing of benefits, rewarding traditional knowledge and its preservation, technology transfer for biodiversity conservation in particular with regard to biotechnology, international access, transfer and use of genetic material, international sharing of genetically modified organisms, intellectual property rights related to wild or domesticated biodiversity and genetic material, and so forth.

In 1993, India's Environment Action Programme emerged as a follow-up to the United Nations Conference on Environment and Development (UNCED) 1992 and Agenda 21. This program strengthened capabilities in environmental assessment, emphasized the importance of environmental awareness and facilitated the process of involvement of NGOs in the tasks of sustainable development. In relation to biodiversity conservation, the action plan set out a program for sustainable generation of non-timber forest produce (NTFPs) and the tasks of afforestation and eco-restoration.

Following the ratification of the CBD, India took more steps in developing new strategies. In 1994, it published a document titled a "Conservation of Biological Diversity in India: An Approach," which set forth the future course of action in biodiversity policy-making. The National Core Group, representing diverse stakeholders, prepared actions and strategies that led to a document entitled "Draft National Policy and Action Strategy on Biological Diversity." In 1996, a new facility was established in the Indian National Gene Bank with long-term storage capacity of nearly 1.5 million samples of seeds and cultures. In 1997, MoEF/UNDP (Ministry of Environment and Forests/United Nations Development Programme) launched the Capacity 21 Programme with an action plan for the period of 1997–2007, which identified economic valuation of biodiversity as a key to sustainable development (MoEF 1998).

In 1999, NGOs loosely coordinated with the government to undertake what was considered the blueprint of the National Biodiversity Strategy and Action Plan. The Biodiversity Conservation Prioritisation Project (BCPP) was India's largest and most comprehensive exercise to prioritize sites, species and strategies for conservation (Acharya 2002). This project supported the development and application of a state-of-the-art methodology for setting biodiversity conservation priorities in a pilot national priority-setting exercise in India. In the BCPP, the Biodiversity Support Program (BSP)[2] worked with an informal consortium of Indian NGOs and research institutions under the direction of a steering group, led by World Wildlife Fund-India (WWF-India), which included representatives from the participating NGOs, the BSP and the government. The project steering group ensured that the process occurred in a transparent and participatory manner and that its participatory priority-setting methodology could be applied in other countries to meet conservation requirements under the CBD. The project produced specific action plans for fifty local sites, analyzed the information from eight focal states to recommend policy changes, and concluded with a national-level workshop to discuss and finalize its recommendations on priority geographic sites, priority species and strategies for conservation. However, it largely failed to influence policy.

CBD norms

Article 6 of the Convention on Biological Diversity requires parties to develop a National Biodiversity Strategy and Action Plan as a roadmap for each country to significantly reduce the rate of loss of biological diversity. The NBSAP would provide the overall framework for national implementation of the three objectives of the convention and be part of an overall sustainable development strategy. The Conference of the Parties (COP) provided initial guidance to parties on the development and implementation of NBSAPs at its second meeting in 1995, which included the suggestions that the process should involve periodic review and improve scientific understanding and socio-environmental assessment. It also stipulated that it should be derived from large participation of stakeholders. The CBD also laid out three critical goals, which the Indian Biodiversity Act, 2002, mirrors, stating that the legislation should "provide for conservation of biodiversity, sustainable use of its components and equitable sharing of benefits arising there from" (Biological Diversity Act 2002, preamble).

The CBD, which is now ratified by 180 countries, requires countries to introduce national policy or legislation on access to genetic resources and benefit-sharing (art. 15 [7]), and to encourage equitable benefit-sharing from the use of related knowledge, innovations and practices of indigenous and local communities (art. 8[j]). Articles 8 and 10 of the CBD require member countries to take strong legal and policy measures to protect the rights, interests and knowledge of indigenous (tribal) and other local communities.

The biodiversity convention also attempts to provide a framework that respects donor countries' sovereign rights over their biological and genetic resources while facilitating access to those resources for users. It requires member states to provide access on "mutually agreed terms" and is subject to the "prior informed consent" of the country of origin of those resources (CBD 1992, art. 15). The convention also stipulates that donor countries of microorganisms, plants or animals used commercially retain the right to obtain a fair share of the benefits derived from such use. This can take the form of both monetary and non-monetary benefits, such as sharing of research, collaboration, participation on product development and access to relevant scientific information on biological diversity. The convention is particularly significant because of its recognition of developing countries' claims to sovereign rights over their biological resources. Second, it emphasizes a new approach to biological resource management, which puts an "increasing emphasis on the potential economic uses of biological resources" (Cullet and Raja 2004, 99) that is particularly significant because of its narrative implications for domestic biodiversity policy, as I will demonstrate in the latter part of this case study.

Article 8(j) in the CBD is entitled "In-situ Conservation," which states, "Subject to its national legislation, respect, preserve and maintain knowledge, innovations and practices of indigenous and local communities embodying traditional lifestyles relevant for the conservation and sustainable use of biological diversity and promote their wider application with the approval and involvement of the holders of such knowledge, innovations and practices and encourage the equitable sharing of the benefits arising from the utilization of such knowledge, innovations and practices" (CBD 1992).

Article 10 is entitled "Sustainable Use of Components of Biological Diversity." Article 10(c) asks signatories to "[p]rotect and encourage customary use of biological resources in accordance with traditional cultural practices that are compatible with conservation or sustainable use requirements" (CBD 1992).

Since the convention entered into force in 1993, the importance of transparency and participation has been underlined. At its first meeting in 1999, the Panel of Experts on Access and Benefit-Sharing (ABS) concluded that "access legislation will only be feasible and implementable if it is developed with the full participation of all those who will be affected by and administering it, such as certain industry sectors, universities, scientific research organizations, ex-situ collections and local and indigenous communities."

The Fifth Conference of the Parties to the CBD in May 2000 called on the Expert Panel on ABS to conduct further work on stakeholder involvement in ABS processes, and it requested the Ad Hoc Open-Ended Working Group on ABS to develop inter alia guidelines on "the roles, responsibilities and participation of stakeholders" (decision V/26). The Bonn guidelines also specified that relevant stakeholders should be consulted in setting up ABS processes, specifically "when determining access, negotiating and implementing mutually agreed terms, and in the sharing of benefits; [and] in the development of a national strategy, policies or regimes on access and benefit-sharing" (Bonn Guidelines 2002, decision VI/24).

The concept of sustainable development which underlies the CBD and is expressed in Agenda 21 very much supports the idea that local people and communities should participate in genetic resource use and benefit-sharing (Glowka et al. 1997). Agenda 21 also underlined that in order to achieve sustainable development, governments would have to incorporate the widest possible participation in decision-making. This has been supplemented by the experiences over the last decades that showed that the most effective policies and laws are usually those that have gained the acceptance of civil society through public participation (Bass et al. 1995). It had been suggested (Swiderska 2001) that involvement of local communities, especially in ABS agreements, could provide resources and incentives for conservation and contribute to livelihoods.

Impact of international agreements

The common interest in biodiversity conservation for India is reflected in the numerous international agreements that India has been party to, including the United Nations Framework Convention on Climate Change (UNFCCC), United Nations Convention to Combat Desertification (UNCCD). Commission on Sustainable Development, World Trade Organization, and International Treaty on Plant Genetic Resources for Food and Agriculture, undertaken both nationally and internationally. Biodiversity as an integrative issue gained attention in the 1980s mainly as a result of three factors: a growing urgency about the unprecedented rate of loss of biodiversity, a greater insight into the different values of biodiversity conservation, and rapid developments in biotechnology (Rajan and Rajan 1997, 155).

The rapid growth of biotechnology is particularly important as it is one of the areas where India wanted to develop its full potential because of its significant biological resources and its strong bases of indigenous knowledge. In India, the three legislative instruments that make up the core of the CBD are particularly important, especially the goals of conservation and sustainable use. This is because by making the exploitation of biodiversity important, it has a significant influence on conservation of those same resources. In addition, the combined impact of the three principles is "an implicit (re) distribution of property rights" (Cullet and Raja 2004, 108). This has been controversial because of the need to redefine ways of articulating private and community property rights – for instance in the case of creating people's biodiversity registers (PBRs). PBRs were a program initiated to manage and organize indigenous knowledge and to integrate it into a broader biodiversity information system. This process revealed many of the inherent tensions underlying participatory processes, including appropriation of traditional knowledge for commercial practices without the just sharing of benefits, threats of over-exploitation and questions of ownership, even at the national level, mirroring some of India's obstacles at the international sphere.

The impact of the domestic legal regime on property rights mirrors rapid changes observed at the international level over the last few decades. The distribution of property rights over biological resources has been a source of tension. After decolonization, states retained sovereignty over natural resources. The Biological Diversity Act 2002 clearly reflects this trend, in retaining a state's sovereignty and yet setting up legal mechanisms for private appropriation of biological resources and knowledge. This law addresses the basic concerns of access to, and collection and utilization of, biological resources and knowledge by foreigners, and the equitable sharing of benefits arising out of such access. The legislation stipulates that a National Biodiversity Authority will "grant approvals for access, subject to conditions, which ensure equitable sharing of benefits" (India and WTO 2000, para. 7). This is in direct opposition to the notion of "common heritage," which the Food and Agriculture Organization (FAO) considers relevant to genetic resources relevant to food and agriculture, as can be seen in its International Undertaking on Plant Genetic Resources for Food and Agriculture set up by the Commission on Plant Genetic Resources in 1983 (Brody 2010). This notion coupled with the "lopsided nature of the IP system, the advances in modern biotechnology and the accompanying spate of bio-piracy grossly disadvantaged indigenous communities, since the implication was to place their resources squarely at the disposal of technologically developed user countries" (Zainol et al. 2011, 12401). The Agreement on Trade-Related Aspects of Intellectual Property Rights (TRIPs) of the World Trade Organization only intensified the issues outlined above. In particular, it attempted to homogenize intellectual property rights (IPR) regimes and constrained countries or communities from choosing its method of conservation and use of biological resources and traditional knowledge. The TRIPS also contained no provisions for protection of the indigenous knowledge, which may not be suited to the safeguards of current IPR regimes. Finally, it has no recognition of the demand to equitably share in the benefits of knowledge related to biodiversity. Instead it heightened inequities that characterize interactions between industrial users of knowledge and communities and countries that contained the bulk of biological knowledge.

These larger trends in trade and intellectual property being pushed by developed countries spurred India's participation in the biodiversity convention. The convention's negotiations provided India with the opportunity to argue for each state's right to choose appropriate intellectual property legislation and to argue against the erection of new barriers to the flow of technology from the North to the South. Before the negotiations for the biodiversity convention began, India discussed its views with other select developing countries in a conference in New Delhi in 1990. Through mutual negotiation, lists of principles that the countries agreed would have to be incorporated into the convention were produced. India argued that while developing countries have the bulk of biodiversity and the responsibility to maintain it, developed countries through technological capabilities enjoyed the economic benefits of the same diversity. Therefore, India pushed for the protection of its own biodiversity and traditional knowledge and demanded just compensation for its use. They pointed out that while formal innovations in the developed world were adequately rewarded, informal ones were not even recognized.

The CBD has two contradicting provisions relating to intellectual property rights. One (art. 16.5) states that contracting parties shall cooperate to ensure that IPRs are "supportive of and do not run counter to its [the CBD's] objectives." This is "subject to national legislation and international law." Another (art. 22) states that the CBD's provisions will not affect rights and obligations of countries to other "existing international agreements, except where the exercise of those rights and obligations would cause a serious damage or threat to biological diversity." These articles allow some maneuverability with regard to IPRs, while recognizing that patents on technology and other IPRs will be subject to international legislation on the same. Thus, countries have to tread a very fine line between the protection of their "public interest" in sectors and rights that contribute to socio-economic advancement. For instance article 8 of the TRIPS agreement does allow for governments to protect national health and nutrition; however, environment, benefit sharing and community rights are given no such waiver, although one could argue that they are equally vital to "public interest." The differences between the articles seem to suggest that while the legislative framework of the CBD does not want to significantly challenge the existing global order, it simultaneously refuses to surrender the politically significant demand by developing countries, of national interest. These inconsistencies in the legal regime have put a greater concentration of powers in the hands of national governments, which civil society has argued have been accompanied by a surrendering of certain community resources to private sector interests (Cullet and Raja 2004).

The CBD is significant for India as the tenor of the legal framework influenced the extent of people's participation in the legislative process. The TRIPS agreement was widely criticized even before its adoption in India, because consultations were not held before ratification. In contrast, the UNDP-Global Environment Facility (GEF) and the Union Environment Ministry's National Biodiversity Strategy and Action Plan was a consultative process that continued for over two years (1999–2001) and included all states and union territories. With the unusual constellation of the government institutions and NGOs, the NBSAP's scale has covered eighteen sub-state ecological zones, ten inter-state ecologically important zones, thirteen thematic-level plans on major areas of biodiversity and thirty reviews of specialized biodiversity areas (Acharya 2002, para. 1). Combined with this has been the process of outreach that included tens of thousands of people including schoolchildren and youth and members of the government administration. Support has been "garnered through biodiversity festivals (*melas*), cycle and bullock-cart rallies, cultural programs, calls for public hearings and participation to put forward their suggestions and concerns of the citizenry" (Acharya 2002, para. 2). India could therefore position itself vis-à-vis the international debates, using the convention as a forum to push for protection of traditional knowledge and informal innovation and showcasing it in the process of a highly visible, highly participative formulation of the National Biodiversity Strategy and Action Plan.

The process for drafting the NBSAP became one of the most "unique and trend-setting" (Anuradha et al. 2001, 20–22). After two decades of gradual decentralization, the ground was ripe for supporting such a large-scale participative movement. However, it ultimately failed to have a uniform effect as a legal document. The inconsistencies of the provisions within the CBD had both an enabling and a constraining effect on the negotiation of the NBSAP plan. The focus on indigenous knowledge and participation in the CBD allowed the system to be opened up to a variety of stakeholders to build awareness about their rights as users and conservers of biodiversity. On the other hand, the national interest that was threatened by the TRIPS agreement, the emphasis on the economic utility of biological resources and the fear of loss of state control over genetic material had serious consequences for the NBSAP process. In addition, the very international nature of the NBSAP that enabled large-scale participation also proved embarrassing for the state when it retreated from the process, as the organizations used this public space to launch critiques on the state's past activities with relation to the governance and management of natural resources.

Linking the global to the local

Domestic actors are more receptive to international norms whose meanings can be negotiated and are framed in a way that they can mean different things to diverse actors at the national or local level. Acharya understands this process of "framing" as the way in which language highlights and dramatizes international norms, in order to make them locally relevant (Acharya 2004, 243). These framing exercises are useful in building linkages between existing norms and emergent norms that are often not obvious and have to be actively constructed, to build salience between international values and local realities. As a result, mediators must work hard to frame their issues in ways that make persuasive connections between existing norms and emergent norms – a process Acharya (2004, 243) calls "grafting." Norms are not monolithic systems of values and ideas which all members, nations or communities are forced to accept. They can exist in different forms, have different nuances and even be associated with conflicting principles that are not consistently adopted by the state in question. States have a possibility to pick and choose the norms most appropriate to them, those that may be consistent with local values, practices and beliefs (Subotic 2009).

There are important processes of exchange between domestic and international norms. Additionally, the filter of domestic structures and divergent norms may produce important interpretations and adaptations of international norms. These interpretations often need important catalysts, in the form of mediators or intermediaries at the domestic level to champion processes that advocate minority or marginalized positions. International norms strengthen the position of mediators at the domestic level and often confer legitimacy on their claims. Thus, according to Finnemore and Sikkink (1998), there is often a "two-level norm game" occurring linking international and domestic demands.

A broader literature review reveals two different mechanisms at work in the process of norm diffusion: one is "bottom-up," and the other "top-down." The former has two paths: the first argues that domestic social actors, even those not linked to broader transnational relationships, exploit the pressure that international norms generate to influence state decision makers. The second illustrates how non-state actors come together with policy networks, at both the national and the transnational level, united in their support for norms; they then mobilize and coerce decision makers to change state policy.

Cortell and Davis point out that this process of interpreting norms does not always involve contestation between international and state actors (1996, 452). Rather, government officials and civil society organizations appeal to international norms in order to achieve their own goals in the national sphere. This is especially viable if international norms are able "to enhance the legitimacy and authority of ... extant institutions and practices, but without fundamentally altering their existing social identity" (Acharya 2004, 248). The role of mediators can be in the form of epistemic communities and international organizations, but also NGOs, think tanks, multinational corporations, and trans-governmental networks (Ban 2010, 21). Recent scholarship (Acharya 2004; Ban 2010; Cortell and Davis 1996) places greater emphasis on the role of domestic actors as these intermediaries who translate and negotiate international norms into locally relevant ideas.

This process is elaborated in the following case study where civil society organizations act as mediators, involved in translating particular norms in locally relevant ways. This study traces the processes of "how domestic political structures and agents condition normative change" – which is concerned with the translation of norms by local actors, rather than a uniform acceptance of norms (Acharya 2004, 240). Very few studies (Acharya 2004) have touched upon the idea of norm localization, especially in the context of India. The translation literature has not yet addressed international norms regarding participation, conservation and sustainability in the context of civil society organizations in a developing country context. In addressing this omission, I show how the norms propagated by the CBD were crucial in fostering deliberation in the policy space in India. Government and civil society organizations subscribed to these norms to further national agendas and translated these norms in locally relevant ways. In the process, we see contestation with the state by civil society organizations such as Kalpavriksh, who highlighted people's rights politics in India and reshaped the terrain of domestic advocacy.

Ratification of the CBD in 1993 did not mean that the Indian state was ready to commit immediately to the introduction of a rights-based internal reform. It did however open the door to multiple opportunities, which became important milestones in rights-based advocacy. The incorporation of the CBD changed the domestic environment in three ways:

1. Framing: the CBD provided a new and "official" way of framing biodiversity issues for the state. The state actively built new frames including ideas of participation, conservation and equity with non-state actors who had been advocating for these rights at different levels. The ratification of the CBD forced state and non-state actors to adapt their rhetoric to the language of stakeholder rights and participation.
2. Leveraging opposition to the TRIPS: the CBD strengthened the assertions of pro-reform NGOs who until then were focused on the negative impacts of patents on genetic resources/traditional knowledge. NGOs took a position that IPRs are not necessarily detrimental as long as India protects its genetic resources. This allowed NGOs to advocate for more participatory, stakeholder involvement in the formation of laws and policies that could extend rights to grassroots innovation and traditional knowledge rather than simply protecting them for patents and from bio-piracy.
3. Strengthening the moral authority of rights-based advocacy organizations: the re-framing of the language along the lines of a rights and participation discourse discredited competing discourses within civil society and increased the legitimacy and authority of advocacy organizations.

The CBD legitimized a new rhetoric of indigenous rights in the area of biodiversity. India committed to three critical goals when it ratified the CBD in 1993: conservation of its biological diversity, sustainable use of its biological resources, and equity in sharing the benefits of such use. It helped to re-define the discursive terrain on which state and civic actors concerned with issues relating to communities and their rights operated. As pointed out above, the CBD allowed for alternative ways of framing to enter the policy arena, one in which participation, voice and control were relevant. In particular, it questioned the culture of "fences and fines"[3] where human use and impact on natural resources are severely controlled. Instead it began to push for the legitimization of access and control of communities over their own resources. The NBSAP went a step further by opening up the policy culture to alternative voices and advocacy measures. In the public arena of civil society, and more specifically within the organizations that specialized in indigenous people's rights, this discursive shift granted moral authority to advocacy organizations. Although they had historically promoted a rights-based approach, they had, until that point, been marginal to the policy process: operating outside the institutional spaces of the state.

"Indigeneity" is a useful conceptual tool for understanding conflicts between indigenous people and nation-states where indigenous groups are powerless minorities. Scholars have suggested that the best way to understand indigeneity was in terms of a "claim to justice" that is "based on awareness of historical injustice the consequences of which have been inherited by contemporary people" (Canessa 2007, 196). This is well placed with the growing understanding of indigeneity as a globalized discourse of rights that are accessed by peoples engaged in local struggles. The CBD, through its focus on resource revitalization, addresses, albeit indirectly, the related aspects of knowledge and culture. It spurred huge investments for the protection of resources, both nationally and internationally. This translated to integrating knowledge systems with more attention being given to the documentation and codification, validation and utilization of biological resources.

Indigenous representatives increasingly speak the language of biodiversity. Support for the CBD's Programme of Work (2004) from indigenous organizations is indicative of this trend. The CBD's Programme of Work sets out a common language and set of objectives. This was very clear from the substantial emphasis indigenous organizations put on Durban outputs and program element 2 of the CBD Programme of Work on Protected Areas (see World Climate Conference 3 2009, resolution 81).

The CBD PoW sets the following as a target:

> The full and effective participation by 2008, of indigenous and local communities, in full respect of their rights and recognition of their responsibilities, consistent with national law and applicable international obligations, and the participation of relevant stakeholders, in the management of existing, and the establishment and management of new, protected areas.
>
> (CBD PoW, VII/28)

It furthermore suggests:

> Implement specific plans and initiatives to effectively involve indigenous and local communities, with respect for their rights consistent with national legislation and applicable international obligations, and stakeholders at all levels of protected areas planning, establishment, governance and management, with particular emphasis on identifying and removing barriers preventing adequate participation.
>
> (CBD PoW, VII/28)

The CBD PoW mentions "full and effective participation" in "full respect of their rights and recognition of their responsibilities" and that such efforts should conform to both "national legislation" and "applicable international obligations" (VII/28). It is clear that the relevant body of jurisprudence would relate particularly to procedural rights given the emphasis on participation, but, with the CBD's emphasis on equitable cost and benefit-sharing, also links it to substantive rights, including, for example, the rights to food and development, along with customary rights to land and water. Indigenous and local communities have particularly highlighted land and resource rights as integral to their customary rights.

NGOs have had considerable influence on developing alliances and creating politically engaged discourses on inclusion (Greene 2009; Martínez Novo 2006). Arturo Escobar (1998, 56) argues that the biodiversity network constituted by global and national "institutional apparatus" promulgates strategies and programs, and "creates knowledge and power based on a techno-scientific idea of biodiversity." While Article 8(j) of the CBD functions to give some attention to local knowledge, Escobar argues that "this attention is insufficient and often misguided to the extent that local knowledge is rarely understood in its own terms or it is re-functionalized to serve the interest of Western-style conservation" (1998, 61). National NGOs, therefore, play a strategic role in adapting international discourses on rights and adapting them to the plans and strategies put forward by nation states for the conservation of biological diversity. If the difference between indigenous and Western knowledge is the ability to legitimize the social construction of particular forms of knowledge (Agarwal 1995), then NGOs and communities can play a valuable role in redefining cultural and ethnic constructions. These social actors situate themselves within the knowledge-power network of the techno-scientific biodiversity discourse, often acting as cultural "brokers" conversing across indigenous and Western worlds, and alternatively resisting, subverting or recreating constructions to serve other ends. This was well reflected in the NBSAP process where the steering committee acted as the brokers translating the contributions of "fisherfolk, peasants, forest dwellers, pastoralists, professionals of various kinds, government officials from different line agencies, NGOs, industrialists, students, even housewives and husbands who have something to contribute" (NBSAP 2000, personal communication, April 18, 2010).

Several people involved in the NBSAP process reinforced the need to reinstate activities that had been traditionally sidelined. Elizabeth Zopari (NBSAP 2000),

NBSAP's media campaign manager, explained the need for using folk media, in addition to more traditional print media campaigns. P. V. Satheesh (Deccan Development Society and member of the TPCG) stressed the "celebration of biodiversity placed in a cultural context" (NBSAP 2000, personal communication, April 18, 2010). The cultural and ecological value, instead of the commercial value of biodiversity, was emphasized through the NBSAP process, encouraging local people to revive rituals and aspects pertaining to the preservation of biodiversity. In particular the NBSAP resurrected the rhetoric of participation, in a culture of centralized planning, backed by the UNDP and GEF. In addition it spearheaded the movement of documenting and collating traditional knowledge, subverting the traditional dichotomy between "scientific" and "traditional knowledge" by creating national biodiversity registers in collaboration with NGO's like Gene Campaign, Research Foundation of Science, Technology and the Environment (RFSTE), the Society for Research and Initiatives for Sustainable Technologies and Institutions (Sristi), Kalpavriksh and the *Beej Bachao Aandolan* (Save the Seeds Campaign). As explained by Madhav Gadgil, member of nodal agency, NBSAP:

> Biological diversity cannot be surveyed by experts alone. The role of people at grassroots level needs to be acknowledged. There is a lot of informal knowledge at field level, which can be collected and then upgraded into formal scientific knowledge for further use.
>
> (D. Das 2011)

With regard to leveraging opposition to the TRIPS, India was one of the most vociferous opponents of revising its patent laws according to the TRIPs Agreement of the WTO, and refused to accept its provisions. NGOs began a campaign of some scale in their protest against TRIPS. Their most effective and forceful argument was that the IPR system, outlined in TRIPs, recognizes only innovations of corporations and does not take into account informal innovations of farmers and communities, especially in developing countries. These NGOs pointed out the negative impact of patents not only on industry, health and prices but also on bio-piracy.

The Earth Summit in Rio de Janeiro in 1992 provided momentum to their protest as the CBD represented the overt shift of developing nations from the common heritage to sovereign control over genetic resources. There was a rethinking of a strategic path to securing gains from intellectual property that would entail extending it to traditional knowledge and grassroots innovation rather than protect it against all individual patents. They applied the ideas of the CBD, arguing that people and communities should have control over their knowledge and derive IPR benefits through it. It would necessarily follow that they then become active stakeholders in the laws or policies arising from biodiversity related issues.

Western techno-centric notions that are biased against indigenous communities bind the protection of intellectual property (Blakeney 1997). Under the existing legal regime, patents are granted for inventions that must be "novel" and assessed by reference to prior technical use, and individual inventors are protected. The systems of knowledge that do not fit this model, including traditional knowledge systems, are denied intellectual property protection. This question of "novelty" has grown to be a bone of contention between countries as the TRIPS Agreement does not clearly define "novelty," leaving members free to define and interpret the terms freely. In the case of traditional knowledge, especially knowledge systems that are undocumented, "novelty" becomes a more ambiguous concept.

Several concerns have been raised at both a national and an international level about the incompatibility between the CBD and the TRIPS developed by the World Trade Organization (WTO). Article 8(j) of the CBD declares:

> Subject to its national legislation, [states should] respect, preserve and main- tain knowledge, innovations and practices of indigenous and local com- munities embodying traditional lifestyles relevant for the conservation and sustainable use of biological diversity and promote their wider application with the approval and involvement of the holders of such knowledge, inno- vations and practices and encourage equitable sharing of the benefits arising from the utilization of such knowledge, innovations and practices.

Article 27.3(b) of the TRIPS agreement states:

> Members may also exclude from patentability plants and animals other than micro-organisms, and essentially biological processes for the production of plants or animals other than non-biological and microbiological processes. However, Members shall provide for the protection of plant varieties either by patents or by an effective *sui generis* system or by any combination thereof.

The CBD negotiations took place after years of North-South conflicts over resources and rising environmental concern. The starting position was that genetic resources were a "common heritage to man" (Downes 1996, 171) and "biodiver- sity information belonged to no one and could be exchanged freely among the countries of the world" (171). Developing countries, justifiably suspicious of the price they will pay for this "free exchange" of information, supported the current view as expressed in article 15 of the CBD that provided national sovereignty over genetic resources, which was tempered by the obligation to facilitate access to genetic resources by other contracting parties.

Over the last few years, through communications to the TRIPS Council of the WTO, countries of the South have emphasized that the rights of the holders of traditional knowledge must be represented and that they must be able to share benefits arising out of innovation based on their knowledge. As McAfee (1999) argues, if market forces determine the distribution of the benefits of biodiversity, then the world's economic elites will benefit disproportionately, further disem- powering and impoverishing the rural poor. At the international level, countries and NGOs have called for a harmonization of the provisions of TRIPS with those of the CBD. In the absence of clear directives in the TRIPS that are in line with members' obligations under the CBD, countries have taken the position that implementation of the TRIPS Agreement may allow for acts of bio-piracy and thus result in systemic conflicts with the convention. Hence, this group of countries has proposed in the WTO that the TRIPS Agreement should be amended in order to provide that members should require that an applicant for a patent relating to biological materials or to traditional knowledge should provide the following, as a condition to acquiring patent rights:

- disclosure of the source and country of origin of the biological resources and of the traditional knowledge used in the invention;
- evidence of prior informed consent through approval of authorities under the relevant national regimes; and
- evidence of fair and equitable benefit sharing under the national regime of the country of origin.

(K. Das 2005)

The dissonance between the CBD and the TRIPS is being negotiated at several levels. While the CBD calls to "respect, preserve and maintain" traditional knowledge of indigenous communities, the TRIPS agreement legitimizes private property rights through intellectual property over life forms. These rights validate the demands of individuals, states and corporations, not indigenous peoples and local communities. The TRIPS agreement resulted in mass protests by indigenous and peasant communities along with NGOs in Asia, Africa and South America (Dawkins 1997). It includes farmers' organizations, indigenous groups, local communities and NGOs, who are fighting at different levels to maintain possession of land and resources and cultures. While their efforts have contributed to the recognition at the Rio Convention that indigenous peoples have used and conserved genetic resources for thousands of years, the CBD, in itself, does not ensure their ownership and management rights to these resources.

Ultimately, compliance with international agreements requires involvement of people within the nation state. Although the implementation of international law may seem to be a top-down process, at the national level the reverse holds true (Bhutani and Kothari 2002). Eventually, agreements at an international level are only as good as their impact on national legislation and movements, and they depend on how domestic governments engage and internalize participation in biodiversity management. This idea of ownership, left ambiguous in the CBD, was taken on by the NBSAP and interpreted as community rights and empowerment through participation, following the idea that the promotion of genuine participation in society is an essential part of a rights-based approach to development.[4] It took on the idea of control of resources as a way to protect and legislate on indigenous resources, strengthening the provisions of the CBD through community awareness and ownership against the influence of TRIPS. In the feedback from workshop participants, during the inaugural workshop, the technical and policy group clarified:

It was felt that an activist streak would help counter negative trends that are taking place. As an example, a campaign against patenting, initiated by a forum like the NBSAP was mentioned, it was however clarified that the NBSAP cannot in itself become an agitational process, but could be a forum for the exchange of views, and a platform to raise critical issues including those linked to WTO and globalization.

(NBSAP 2000, personal communication, April 18, 2010).

The issue area of biodiversity introduced norms brought into the limelight by the CBD. Non-state actors had an important role in mobilizing to link the global to the local: in translating the norms into a language and concept that could be easily grasped by all. According to one legal activist, the Hindi equivalent of the concept of biodiversity is "jaiv-vividhata,"[5] which has no colloquial equivalent, and translating the concept of an integrated system of biodiversity at different local levels proves difficult. The second challenge was to match these norms to preexisting cultural and political domestic structures, which would include legal structures and mechanisms of decentralization. One of the typical demands that activists made was for the provisions of the CBD to be incorporated into domestic law, since this was seen as an important step towards making rights transparent and equitable. In addition, the planning process outlined by the CBD stressed that stakeholder participation is critically important and the coalition involved in formulating the NBSAP pushed for precisely this model of participatory decentralization.

The formulation of the National Biodiversity Strategy and Action Plan, 1999, was praised for drawing consultations from various sectors, involving a high level of informed debate and including unprecedented scales of participation in a political space that had traditionally subscribed to narrow notions of governance. This level of participation guides the focus of policy analysis to the meanings (values, beliefs) that policies embody for their multiple stakeholders and the ways in which those meanings are communicated or narrated in a particular cultural and political context. In order to study the diffusion of norms and understand what forms they took in a national context, we must focus on the main ideas that these

Table 3.2 Convention on Biological Diversity (CBD) (source: Anderson 2008,144)

Core norms	• Conservation, sustainable use • Participatory planning process • Access benefit sharing
Secondary norms	• Access to genetic resources • Recognition of traditional knowledge and indigenous people's rights • Capacity building
Core rules to be implemented by governments	• Domestic policy and legal measures to ensure equitable sharing and technology transfer • National strategies, plans for programs for conservation of biological diversity and its integration across sectors • Access to genetic material is subject to national legislation; contracting parties shall create conditions for facilitated access to resources

norms transmitted at the international level. The norms that governed these interactions are set out in Table 3.2.

Deliberations and participation in the National Biodiversity Strategy and Action Plan

Institutional deliberation

The foray into biodiversity planning had its basis in informed dialogue but retained
the top-heavy structure that characterizes policy-making in India. In 1994, the
Ministry of Environment and Forests (MoEF) was investigating the requirements
of a biodiversity law and held consultations with representatives from government
agencies, NGOs and academics to discuss the need for a national action plan on
biodiversity. Initially, a core group consisted of representatives from various government authorities and autonomous institutions under the Government of India
dealing with some specific issues. These included sustainable utilization, public
participation and benefit sharing, identification of relevant technologies and the
role of industry in the transfer of technology under the CBD, amongst others.

A series of six consultative meetings was held for the core groups within which
each sub-group was mandated with drafting an action plan in its area of expertise.
The framework they drafted was sent to various scientific institutions for comments and inputs. The core group, along with its sub-groups, deliberated over the
revised draft. It was then decided that a "nucleus team" – comprised of the Forest Survey of India, Botanical Survey of India, Zoological Survey of India, and
Wildlife Institute of India, under the coordination of the Forest Survey of India –
would finalize the form of the main document. These educational and research
institutions remained firmly subordinate to the ministry and thus compromised the
transparency and reach of the discussion on biodiversity. The team held further
discussions with scientists and NGO representatives, and it came up with a new
draft in June 1995. This draft was severely criticized for not taking into account
the discussions and intensive debate that had been integral to the action planning
process (Taneja and Kothari 2002).

In June 1997, a MoEF consultation on the action plan evaluated ongoing strategies and programs and assessed current and future biodiversity conservation and
sustainable use needs. Input was sought on eleven subject areas, which are covered in the macro strategy document:

- legal and policy framework;
- survey of biodiversity and a national database;
- in situ conservation;
- ex situ conservation;
- sustainable utilization;
- indigenous knowledge systems;
- innovations and practices and benefit sharing;
- peoples' participation;

- institutional framework and capacity building;
- education, training and extension; and
- research and development activities and international cooperation.

(Taneja and Kothari 2002)

This macro strategy highlighted links between the different levels of government, civil society and local bodies that were critical to effectively conserve and utilize biodiversity.

It was also decided "scientific examples, illustrations and case studies should be prepared for assessing the value of biodiversity, both economically and culturally" (Taneja and Kothari 2002, 379). Compiling biodiversity registers of indigenous knowledge was suggested after detailed study and discussion of the implications of such documentation. It was concluded that all activities of evaluating and monitoring of biodiversity and implementation of conservation must focus on the local level, with national-level institutions playing the role of supervisor.

In October 1996, the MoEF, through a liaison with the UNDP, applied to the GEF for funds to prepare an action plan on biodiversity. The project was approved in March 1999, and the original plan preparation process was reviewed. It was suggested that this expert-led centralized structure be replaced with a more decentralized arrangement. Other suggestions for the planning exercise included increasing participation, discussing the roles and limitations of central and state governments and ministries, developing strategies for grassroots conservation and awareness, and considering the larger socio-economic and fiscal dimensions of biodiversity conservation (Taneja and Kothari 2002).

Against this backdrop, the MoEF evolved a mechanism for the preparation of the NBSAP that would take it beyond a review exercise at the central level. It was decided that coordinating the preparation of the NBSAP would be given over to an external agency. Proposals were solicited by MoEF, and, of the six agencies that responded, Kalpavriksh (an NGO based in Delhi and Pune) was chosen. Administration of the project was entrusted to Biotech Consortium India Limited (BCIL), a secondary research organization based in Delhi that deals with matters of environmental engineering and biotechnology. Kalpavriksh expanded the institutional arrangement for preparation of the action plan to include a fifteen-member technical and policy core group (TPCG). The members of this group include "experts" from different parts of the country and different sectors of work related to biodiversity and conservation. The time for developing the NBSAP was set at two years, and it began in February 2000, with the first meeting of a Steering Committee at the national level (Taneja and Kothari 2002).

MoEF continued to be the main executing agency of the NBSAP, and the budget for the entire process was approximately Rs. 4 crores (US$916,588), with Kalpavriksh receiving a small part and the rest going to other partners of the core group. The implementation of the NBSAP project was overseen by the Steering Committee that was constituted under the chairmanship of Vinod Vaish, special secretary, with R. H. Khwaja, joint secretary of the MoEF. The other members of the Steering Committee were the additional inspector general of forests (wildlife);

representatives from the Department of Agricultural Research and Education, Department of Biotechnology, Department of Science and Technology, Department of Development, Department of Indian System of Medicine and Homoeopathy, Ministry of Social Justice and Empowerment, Department of Economic Affairs, Planning Commission, UNDP; and some additional experts. This committee met once in every six months to be updated on the process of the NBSAP, approve key decisions and provide guidance.

The function of the TPCG was to coordinate and conceptualize the preparation of the NBSAP. In conjunction with the MoEF and its own network, the TPCG identified nodal agencies to prepare biodiversity action plans at state, sub-state and eco-regional levels. Each nodal agency (or working group) prepared a strategy and action plan (SAP) covering the theme for which they were responsible. These SAPs were then collated with the final responsibility remaining with the TPCG. Apte (2005) points out that these coordinating agencies had to be chosen with care with a specific set of criteria – namely, relatively independent and neutral, not too overtly favoring either the agenda of livelihoods or the conservation, and acceptable to both government and non-governmental organizations. In addition, the perceptions of coordinating agencies would influence stakeholders and ultimately the outcome of inclusive participation. The TPCG found that their "neutrality" played a significant role and they organized a process as neutral as possible in order to avoid a capturing of interests.

In order to create a forum to provide orientation and guidance to coordinating agencies, and subsequently to discuss inputs and the progress of the NBSAP, three national-level and five regional-level workshops were held by the TPCG. The national-level workshops were held in Delhi. The regional workshops were held at five different regions around the country, for the coordinating agencies of those regions to gather. The workshops were open to anyone who wanted to speak. They were structured forums to discuss strategies that could be adapted to the regional context and provide networking opportunities for organizations in the same region.

Advocating for participation

Participation was built into the process of the NBSAP: NGOs in partnership with the government attempted to make the process both large scale and representative. For the TPCG the process of creating the NBSAP was as important as the plan itself. Implementation was not a mandate of the formulation process. That would come later through specific government action. In order to keep the process as open and inclusive as possible, all working documents and minutes of the TPCG were accessible on a public website until January 2002. The TPCG also encouraged coordinating agencies to record their conflict of interests and strategies to overcome them and that these should be incorporated into the plan. This was in keeping with the vision of the NBSAP, which was to "create a decentralised planning process that would result in a plan that would carry within it the priorities and aspirations of the common people of India" (Apte 2005, 15).

Table 3.3 Methodological notes in NBSAP guidelines

NBSAP emphasis on participation

It is critical that in all these activities, there be maximum participation of all sectors (governmental agencies, local communities, independent experts, private sector, armed forces, politicians, etc.), especially through:

1. Making the process of working fully transparent;
2. Inviting public inputs at every step;
3. Making all relevant information available to the public;
4. Using local languages in all key documents and events;
5. Respecting the output of "lower" level (e.g. sub-state) BSAPs and information, and integrating them into "higher" level (e.g. state and national) BSAPs; and
6. Allowing for a diversity of opinions and approaches to be reflected in the process and in the final BSAPs.

Source: Methodological Notes in NBSAP Guidelines Concept Papers (MoEF 2000), distributed to all coordinating agencies

Early in the process the TPCG produced a brochure: "call for participation" (CFP) and advertisements which were disseminated in seventeen regional languages. "The last page of the CFP was a 'cut-out' coupon for people to fill and send in. The coupon indicated five options for participation (organising local meetings; holding an inter-departmental meeting; sending in existing information or documents; contributing new written material; co-ordinating the preparation of an action plan in the respondent's region). A sixth option was any other mode of participation that could be suggested by the respondent" (CFP brochure; in Apte 2005, 98). Approximately 30,000 copies were distributed countrywide. Many who responded became community coordinators and members of the state coordinating agencies and local advisory committees (Apte 2005).

A national media campaign was developed at several levels – nationally and at a state and sub-state level. The latter were also encouraged to develop indigenous tools and strategies in order to maximize "cultural matching." For instance several coordinating agencies independently revised and adapted the CFP to match local contexts and communication styles: for example in coastal Andhra, where a simple, short, locally adapted version was produced. Although the CFP was originally written in English, the TPCG invested a lot of time finding translators to convey the messages of the more technical brochure to simple, local adaptations. This was challenging, and several interviews conducted by the TPCG were revealing:

> [I]n Andhra Pradesh and Karnataka people felt that the translations were too literal and the language used too difficult; it could not be easily understood by laypersons. In addition, people need simple, practical write-ups, in simple language, with a local flavor – and not general prescriptions or theory of it from Delhi. They would not have the time or patience to understand it.
>
> (Apte 2005, 100)

At the national level, the media campaign aimed at creating awareness and momentum through the use of print media, newsletters, national television spots, radio spots and magazine articles: *Folio*, the magazine supplement of a national newspaper, *The Hindu*; and *Chandamama* and *Vasudha*, which are children's magazines. In addition, in the last year of the NBSAP process, four journalists were named "media fellows," and around twenty-five articles were written in both Hindi and English. There were also three thematic workshops planned in conjunction with other grassroots organizations with tribal communities, farmers and wildlife conservationists in Delhi, Andhra Pradesh and Karnataka.

The media strategies at the local level had to be adapted in several ways; for instance, the scope of communication was limited to the area where the BSAP was being prepared. Language had to be simplified as seen with the CFP, and even the "look" of the disseminated articles had to be adapted in order for it to be made interesting for local consumers. An interesting observation by a TPCG member was that the material used to make the posters communicated in different ways to urban and rural audiences: "For the cities we went in for things that were textured and had a handmade look about them, and which are ethnic looking. But at a village level, people prefer things that are machine-made, and that offers a look of credibility and quality to them" (Apte 2005, 102).

As supporting measures at the local level, mobile biodiversity festivals were organized to gain inputs from the grassroots. This was particularly helpful in the Deccan sub-state plan, where more than 75 villages and 400 people participated. In addition public hearings and village-level consultations were conducted in some states; these were more unstructured forums to elicit the views of more marginalized voices.

Conflicts and failure

Storylines in civil society

The role of civil society as an advocate of particular marginalized voices in a deliberative system is explained here. It demonstrates the idea of a "deliberative continuum" where civil society utilizes structured professional forums to thrash out its strategies of advocacy as well as unstructured public forums to test and debate the same.

The role of advocacy within the NBSAP was twofold. The first role of advocacy was to communicate the message of the process, building ownership and participation at the state and sub-state level. The second role of advocacy was to bring ownership at the state and policy level, in legitimizing a particular vision of rights-based advocacy.

The core policy that the CBD builds on is that environmental concerns and equity should take precedence over economic concerns. This has been a definite source of tension between the CBD and the WTO (Rosendal 2000). In an Indian context, this is also a conflict between the state and different interest groups. The dominant discourse (of the state) has to be interrogated against the different storylines and discourses propagated by the different actors and confront the inherent tensions between them.

State narrative – ecological modernization

The discourse of ecological modernization, in the context of biodiversity, carries a neo-liberal rationale, emphasizing the economic and market aspects of biodiversity by putting an economic value on ecosystem services. Hajer (1995) indicates that "ecological modernization" forms the underlying structure of the CBD. He underlines that this is a discourse that "does not call for any structural change, but is, in this respect, basically a modernist and technocratic approach to the environment that suggests that there is a techno-institutional fix for present problems" (32). On the one hand the convention underlines the principle of conservation of the biosphere; on the other it also sets forth a framework for wider appropriation. In the 1990s, leading environmental institutions, along with the World Bank, inducted this discourse into policy papers, elevating industry and international environmental agencies as main actors in the preservation of biological diversity (Flinter 1997, 148). This created an overriding discourse that is clearly biased towards market considerations and suggests that the loss of biodiversity is a result of inadequate management by developing countries. This is echoed by the chairman of the National Biodiversity Authority of India, who writes: "The time has come for India to link economic indicators with ecological indicators to deal with development. If we consider 'nature' as an asset, we need to do something to secure this for our own futures if not for the next generation" (Pisupati 2012, last para.). He points to the need for appropriate policies that are "taking a precautionary approach," in terms of conservation, valuing "public goods" and "realistic markets" and the role of governments in enhancing private investments in ecosystem-based markets.

Out of the CBD's three main objectives, two directly relate to economic and distributional issues, and its preamble, which states that biological diversity is the "common concern of humankind," implies a global responsibility. Thus, while stating sovereignty over biodiversity within their jurisdiction, there is an additional implication that all biodiversity is the common heritage of humankind (Miller 1998, 181). The objectives and statements about sovereignty imply a movement from an open-access regime to one where access is structured and regulated within the boundaries of a state and is determined not just by intrinsic value (conservation) but also market values (sustainable use) and the consequent distribution of benefits (fair and equitable sharing). As Flitner (1997) puts it, the new legal framework can be seen as a materialization of some of the central elements of the biodiversity discourse which "pretends a positive correlation among the conservation of biodiversity, the growth of the bio-tech industry, the accelarization of capitalization and integration into the world market of 'traditional societies' with their 'undervalued resources'" (156). This was well reflected in the criticism to the Biodiversity Act, 2002, which was passed before the NBSAP process was completed. Though one of the objectives of the act is explicitly stated as the recognition of local rights to biodiversity, the act has come under criticism for vesting most of its power with national and state biodiversity boards (Srinivas 2000; Kalpavriksh n.d.). Moreover, others have criticized it for primarily addressing the concerns of industry (Srinivas 2000). Although there are stipulations for the set-up of local Biodiversity Management Committees (BMCs), the committees remain mostly managerial in nature with serious gaps as to what the rights of these committees are. Others, like Suman Sahai, have pointed out:

> The area on which the biodiversity act is silent is the very area at the centre of a raging global controversy, in which Indian civil society has been very vocal in protecting local communities from the damage inflicted by patents on biological resources and indigenous knowledge.
>
> (Devraj 2002, para. 4)

Sovereignty may be said to delineate the independence and power of a state, with a "government, a people, and a defined territory to do everything necessary to govern itself, including the making, the execution, and the application of laws, in both civil and criminal matters, free of unwarranted interference from other states or foreign actors; it is a state's exclusive right of control over persons, things and activities in its territory" (Brahm 2004; quoted in Zainol et al. 2011, 12403). This sovereignty determines a state's jurisdiction, or the ability and power of a state to make, adjudicate and enforce laws, and is also territorially limited (Blakesley 1999). It is felt however that

> [t]he CBD principle of national sovereignty, by which India claimed a right to its independent management of its biodiversity within the global framework, must further translate into community sovereignty for truly local level decision-making on resources and for application of local know-how.
>
> (Kohli 2007, para. 11)

In contrast to this, it was clear through the process that the government's idea was that the national government has rights over biodiversity.

The Ministry of Environment and Forests (MoEF), during the NBSAP process, saw itself in a facilitative role. It invited the WWF-India, Biotech Consortium India Limited (BCIL) and Kalpavriksh to make presentations to carry out a process that would involve as many stakeholders as possible. The Department of Biotechnology promoted BCIL, but the MoEF liked Kalpavriksh's profile. BCIL was then given the role of financial, administrative and technical management. The role of BCIL, which was conceptualized as a science-industry consortium for biotechnology, had been termed "unreactive" through the process. However, it was stressed that the BCIL only has links to the government and is not controlled by it.

The process of the NBSAP started out with full ownership from the government, and many government officials and bureaucrats, at different levels, remained engaged with the process even after it had formally been tabled. Thus, the state is in no way unified or a monolith. However, a common discourse of neoliberalism has been traced, structuring the policies of the state especially towards marginalized groups. This manifests itself in its ambivalence towards intellectual property rights, in its slow annexation of common property resources and through the idea of fences and fines where states hold forest land in public trust for the purposes of conservation. Theorists have argued (Rudolph and Rudolph 1987) that besides labor and capital, the Indian state itself is the "third actor," which, though constrained by conflicting demands from different actor groups, has sufficient political authority and economic resources to carve out its own agendas. This means the Indian state also selectively chooses agendas to give support to and remains ambiguous about others. Many in civil society see the state strategically pushing the interests of the strong and neglecting the demands of the weak. This is underlined by Abhijit Sen[6]: "It is true that the growth process has been largely unequal but in democracies governments also respond to where financial resources lie. The important thing is that people speak up." This selectivity also holds true for participatory processes as R. Sreedhar of Mines, Minerals and People (personal communication, April 24, 2010) notes: "Participation is being selectively used. Introduction of BT cotton was made participatory but green tribunal bill was hush hush, though it was more expansive and affecting a larger number of people."

In addition, ecological modernization adopts an expert-led, science-based policy framework (Hajer 1995), where scientists and experts are given the role for defining and finding solutions to environmental problems. The lack of "scientific tenor" of the NBSAP document created by the TPCG was cited as one of the recurring reasons for failure and supports the state's stance on ecological modernization. Positivist science-driven inputs continue to provide a powerful discourse for justifying regulatory responses and therefore the boundaries of political decision-making (Hajer 1995; Fischer 2003). Within this approach, policy interventions have to be legitimized by scientific evidence (Hajer 1995; Fischer 2003), and therefore viable alternatives framed by civil society or oppositional politics need to be framed within a scientific discourse in order to the palatable to the state. Hajer (1995) points to the growing importance of the "civil legitimacy" of scientific research or "socially acceptable science" in democratic environmental governance, where a range of voices have rights in participating in policy-making processes.

Civic environmentalism – the narrative of equity

Economists focus on efficiency, but many of the questions in the environment arena are ethical by nature, and it is by no means evident that questions of efficiency should dominate environmental decision-making (Johansson-Stenman 1998). The underlying argument of civic environmentalist is environmental justice, sustainability, equity, participation, primacy of local knowledge systems and emphasis on inclusion of local stakeholders (Bäckstrand and Lövbrand 2006; Arts et al. 2010). It often encompasses a synergistic role between government and civil society that takes the form of state officials enlisting the help of civil society to provide support to the activities and programs run by the state, and social actors pressing the government to implement policies that encourage the expansion of social action.

International environmental treaties are usually agreements among governments binding them to change aspects of behavior within their borders. However, without active civic engagement, governments cannot actively undertake such agreements. Within the state, information needed for effective environmental decision-making is often geographically dispersed and site-specific. And governments often do not have the administrative capacity to collect the necessary data. An MoEF official substantiates this saying that the involvement of an NGO in the NBSAP process was a "good opportunity to use the process to create plans at a state and sub-state level" (personal communication, April 19, 2010).

The core ideas advocated in the process on the NBSAP, gleaned through interviews of the TPCG, were that development in India should be people centered in that people should control India's natural resources. This belief was based on the shared experiences of centuries of colonialism and heavy-handed bureaucracy. Many interviewed viewed their role in the coalition as a service to their country, for the greater good. There was also a general belief that new biotechnologies would rapidly spur demand for access to genetic resources and those communities who owned this knowledge and resource would be exploited. The main idea was to secure democratic control and widespread knowledge about the value of these resources. This also lead to a belief in participation while formulating laws or policies that affected various stakeholders and the central belief that their ideas and voices had to be reflected in any policy that hoped to be relevant.

"'Activists' often saw participation in the network as a temporally limited, yet important opportunity to make nuanced critiques of the state, the market, and at times, the other groups collaborating with them" (Anand 2006, 479). They used the space provided by NBSAP process to launch directed and sophisticated state critiques, including critiques of its agricultural development policies and its command-and-control forest policy. Sections of the plan alluded to authoritarian conservation paradigms of the state, underlying the inequity in policies that directly impacted subsistence livelihoods. The NBSAP also pointed out to institutional structures that destroyed biodiversity through "root causes," such as unequal trading regimes and contracts (see MoEF 2004). "A significant section of the [TPCG's] report highlighted the importance of local community rights to biodiversity and the dangers of an intellectual property rights regime that privileges corporate knowledge and private ownership (patenting). In terms of strategies, the report specified that 'empowered local community institutions' should be the implementers of the plan" (Kothari 2004, 4724). In reaction, MoEF, at several points in the process, asked the TPCG to "tone down," insisting that it follow the tenor of a policy document and not one of activism. Ultimately they dismissed the TPCGs attempt at creating an NBSAP as "controversial" and "messy" saying that "the government has to be careful about every word in a policy" and that "in hindsight that is the problem of involving NGOs" (MoEF official, personal communication, April 19, 2010).

Interviews with people responsible for drafting the NBSAP reveal certain narratives that cement these coalitions together. The fundamentals of this narrative within the network drafting the NBSAP is best revealed by a member of the Wildlife Conservation Society and of the TPCG who identifies an underlying unity to the network and not "diversity for [diversity's] sake." They were bound by certain values such as the following: "Pro-participatory planning processes, primacy to local communities, traditional knowledge and a 'healthy disrespect' for big corporations or big governance" (TPCG member, personal communication, July 18, 2009). It is important to note that conservationists are traditionally a very strong group in India's environmental policy landscape. They were not co-opted during the processes of biodiversity policy formulation and in fact remained wary of the ideology and movement underpinning the process.

Several people got involved in the process merely to filter the discourse of equity, and build ownership amongst people towards the resources in their stewardship. According to Ashish Kothari, the idea was to build empowerment through mass participation:

> The cornerstone of the NBSAP process is participation. The involvement of a range of people has been seen to be so important that the TPCG and the MoEF have spent several months simply planning the process and identifying several hundred people from diverse sectors. The TPCG has prepared and printed about 25 methodology notes and concept papers to guide the participating agencies.
>
> (Kothari 2001, para. 22)

The process began with the hope that it would spur "powerful people's movements[,] and NGOs are likely to be able to use the biodiversity action plans as tools for change" (Kothari 2001, para. 27). At the outset it was believed that

> [i]t will be hard for the government or other agencies to ignore the work, needs and aspirations of thousands of individuals and groups, and if they still do, hopefully they would mount serious pressure to force the NBSAP implementation. At the very least, the NBSAP process will lead to a nation-wide churning of ideas, fresh ways to visualizing the society and its relations with nature, an in-depth questioning of developmental and economic dogmas and most important how to transform centralized, top-heavy planning processes into truly participatory ground-up ones.
>
> (Mathur and Rajvanshi 2001, 13)

In analyzing state receptivity to "policy innovation," one has to also document the frames in which a state is most receptive to ideas. The challenge of this policy was the tension between "scientific framing" and "lay language." The TPCG attempted to adapt the ideas expressed in colloquial language gleaned from the process to the language of policy that lead to dissonance in the tenor of the policy documents. In these particular policies, deliberation played a crucial role in resurrecting marginalized discourse. It employed lay knowledge, framing it within scientific

Table 3.4 Narrative conflicts in the NBSAP process

Civic environmentalism	Ecological modernization
Comments on the process by various TPCG members, personal communication, May 2010–July 2010 "As government people were involved it was assumed that the NBSAP would become policy. There were people from the government sitting in right through and technically approving the process." "Biodiversity is not seen as technical because we talk at the end of the day about natural resources. Somebody or the other is benefiting or not. This automatically involves the rights of people for using these resources. We need to see how resources are governed in our country and for who?" "The network debated development vs. conservation. The TPCG represented a mix of that. Debate resulted in getting people somewhere at the centre of that debate 'because everyone may not be happy with someone who is too much of a wildlife extremist' or someone too pro 'industry funded projects.'" "Colonial forms of government reflected in attitude of bureaucracy . . ." "Dialogue between NGO and government closing up . . ." "Forest department far more oppressive than Zamindars . . ." (landowners). "The government agenda is neoliberal development. Access Benefit sharing hasn't reached the village level. No one knows about it. It can come once people have control over their resources." "Overlap between minerals, biodiversity, poverty, people and mining . . ." "The draft NBSAP came up at a time when MoEF was busy fitting its agenda to India's economic development priorities." "Policies are best not to be too detailed. They have to be seen as guidelines."	**MoEF official, personal communication, April 19, 2010** "Some people within the ministry were crucial for the Indian Biodiversity Act to be brought to force." "Ngo's bring up certain issues as it is their bread and butter to bring them up. They have to keep raising the issues. The Environment has its own administrative structure. To make a policy maker understand issues is a subjective issue. The centralized nature of the environment administrative system has advantages and disadvantages." **MOEF response to "Short Notice Question" raised by a member of Parliament, 2004** "The draft report which has been prepared by the organization Kalpavriksh headed by Mr. Ashish Kothari is a consultant's report which needs to be reviewed at the government level for financial, administrative and legal implications and for scientific accuracy. A consultant's report cannot be accepted as a National Plan unless it is vetted and endorsed by the government. The draft report contains numerous irreconcilable discrepancies, scientific inaccuracies, implausible and unacceptable recommendations, which if published unedited may put the government to great embarrassment and invite international ridicule and criticism." "Scientifically inaccurate . . ." **Interviewee, TERI, June 25, 2009** "In a country document of international significance highlighting only the negative processes. No government can accept it. This was related to conceptual issues as well." **Member, Karnataka State Biodiversity and Action Plan, personal communication, June 25, 2009** "Biodiversity remains politically small fry, thus government remains poor in paying attention to these areas as opposed to commerce or finance. NGOs have to keep raising issues as it is their bread and butter. However

Kanch Kohli, TPCG, India Together, March 2009

"[T]he word bio-prospecting is embedded in the language of access for the purpose of trade, which is what the MoEF has in store for India's biodiversity."

Kalpavriksh, Environmental Action Group, Background paper to the NBSAP (n.d.)

"All of this needs to be viewed in the light of the changing face of environmental governance in the country which gives primacy to economic growth and investment over environment, conservation or people's livelihoods."

there is a serious lacuna as there are no concrete plans. One cannot remain in the position of barking dogs."

"In a country document, that reflects only the negative processes. No government can accept it."

"Even though the consultancy was given to Kalpavriksh, it does not mean that the ministry has to agree 'word to word.'"

Member, TPCG, personal communication, June 27, 2009

"Main peculiarity was that people assumed that it would be the main NBSAP document."

"Ministry is democratically answerable. TPCG is not. However, even TPCG had to be approved by the ministry."

"Choosing experts brings into question how the TPCG was chosen. But you often choose experts who think the way you do. So who actually wields the power? It's much less nebulous in the political system. It would depend on who the bureaucrats promote or boils down to the political person's skill and abilities and knowledge of the subject and reach."

Consultant, GEF, personal communication, April 15, 2009

"Policy is made by people who are really experienced. When I was out of the ministry, I thought they were useless but when I started working with them I realised how experienced they are and willing to listen to logic. Civil society does not always have to criticise government in order to prove that they are civil society."

methodology to put the issue of biodiversity on the political agenda. Lay knowledge is defined here as knowledge that is local, "nonscientific," "hard earned," "less formally organized" and related to "self-identity" (Michael 1992, 323). Eden (1998, 426) characterizes this form of knowledge as incorporating "extended facts" including "beliefs, feelings and anecdotes." These different constructions of knowledge are alternately and opportunistically used in activist strategies and campaigns and in state-initiated deliberative processes for purposes of advocacy and placing issues on the political agenda (Scott and Barnett 2009). The conflict often lies in the space that policy is willing to give to these discourses. In the case of the NBSAP, the state perceived the strategies of civic environmentalists as "scientifically unsound," unable to reconcile the different demands on policy-making.

The deliberative failure of the NBSAP

The decision to recruit an NGO like Kalpavriksh, who had been working in environment and social issues since 1979, was revolutionary. This was done to generate independent information about the state and scope of biodiversity in the country. It would also evaluate the state's progress in designing new policies that would be in line with the CBD specifications. The influence of alliances and working groups like the TPCG depends ultimately on the strength of the non-state actors, the strategies they choose to adopt, the ownership they manage to build, the institutional matrix in which they operate and political strategizing (Grugel and Enrique 2010). In India the TPCG brought together a diverse range of people who had common beliefs in participatory values but very little experience with a project of this scale and a lack of political strategy. Initially in the case of the NBSAP, there was an ownership from the ministry with its senior bureaucrats inviting scrutiny into their planning processes.

In the inaugural workshop of the NBSAP, Dr. Ashok Kundra, special secretary in the Ministry of Environment and Forests (personal communication, NBSAP, 2000), stated, "Biodiversity is an interdisciplinary subject and encompasses a whole range of issues and activities which would require commitment of all stakeholders." He expressed confidence that the project "will generate a number of schemes with the active involvement of state governments and relevant stakeholders including representatives of village communities" and noted that "a broad spectrum of biodiversity stakeholders ranging from policy makers to academicians, subject matter specialists, grassroots level workers and media are participating." He went on to comment, "[D]eliberations were sure to provide a practical framework for formulating State and local level plans for conservation and sustainable use of biodiversity." In 2001, India's second report to the CBD stated unequivocal support for the process, terming it "India's biggest planning and development process" (MoEF 2009, 18). However, in 2004, the report undertaken by Kalpavriksh was demoted to a technical report leading to speculation and criticism:

> Throughout the four-year process, our mutual understanding was that after going through full consultations and peer reviews, the final report we present

would be the final action plan. All previous versions of the report were in fact released under the MoEF's name as the 'draft' action plan, not as a draft technical report. Each of these versions was subjected to widespread public review; a chance was given to all State governments and all relevant Government of India Ministries to comment. Our final report was based on all these inputs and also on a page-by-page reading by MoEF scientists. Why the MoEF had to repeat the process of peer review after once having gone through all this is not clear; one can only surmise that it was looking for a way to delay the final report.

(Kothari; quoted by Bavadam 2006, para. 5)

The bureaucracy and some officials within the Ministry of Environment and Forests (MoEF) in general were wary of accepting unequivocal participation and rights-based claims. As a group, they were far less critical of the policies that focused primarily on conservation, as conservationists have not undercut their essential role as the managers and guardians of natural resources. India's ratification of the CBD provided a difficulty in reconciling top-down planning processes with participatory policy deliberation, especially as the emphasis in the NBSAP was on community-led development, full participation of key stakeholders in decision-making and widespread decentralization. Though parts of the ministry were involved with the large-scale formulation process of the NBSAP, in the end, they could not allow the opening up of the system to a degree that would have resulted in a significant loss of control.

From a promising beginning, the ministry that had showcased the report in international forums silently withdrew its support in 2004. The new administration had ideological differences with the idea of NGO participation in consultations with the ministry. Its objections were particularly strong in relation to "radical NGOs." The MoEF released a circular titled "Guidelines for selection of non-officials to various statutory, judicial, and other committees and other bodies in MOEF." The rationale as stated was:

In the past, inadequate attention has been paid to the question of whether the non-official persons considered for appointment to these bodies have the relevant background or experience for the role. As a result, the decisions and/or advice received from these bodies may have been sub-optimal. Accordingly, it is advisable to prepare and adopt Guidelines for such appointments.[7]

This circular contained clear definitions of "Non-Official" and "Professional Voluntary Organization" and set out strict guidelines on selection, age restrictions, membership and participation in decision-making bodies. This circular was later withdrawn in 2009. Many saw the period of 2004–09 as a regressive one for the MoEF where there was a structural weakening of civil society contributions to the ministry's programs and a dilution of democratic process.

With a change of secretary in the ministry, the process was completely stalled until, in 2005, Kalpavriksh released the draft report that had been prepared in the four-year-process as the "people's plan." A day after the report was made public, the MoEF issued a press release stating that a group of scientists appointed by it had called the report "scientifically invalid."

However, prior to its publication, a peer group whose suggestions had been incorporated in the final version had reviewed the technical report. A MoEF official had also sat with the drafting team for over two days going over the pre-final draft. Moreover, the MoEF itself with the ministry's official name on the copies had publicly circulated all earlier versions of the document titled "draft NBSAP."

In 2004, in one of the responses to the parliament's question, MoEF stated that the final technical report had several factual inaccuracies. According to an MoEF official (personal communication, April 19, 2010), "there were sensational statements, not scientifically substantiated." Examples given included the following:

- "India's model of development is inherently unsustainable and destructive of biodiversity. . . . [I]t needs a drastic re-orientation."
- "In India, a number of biodiversity elements have been subjected to impacts of inappropriate trade systems. . . . [I]mpacts to biodiversity from trade are likely to significantly increase in the next few years, with India's acceding to the World Trade Organisation's treaties."
- "India has played an inadequate role in advocating conservation and sustainable use of shared resources with neighbouring countries at South Asian fora such as SAARC [South Asian Association for Regional Cooperation]."
- "There has been inadequate use of international human rights treaties and fora by India to promote the cause of biodiversity and livelihood security."

(Badavan 2006, para. 5)

It took another two years for the ministry to release its draft plan authored by two officials from within it, which finally came into force as the official National Biodiversity Plan in 2009. In August 2007, the MoEF released its draft National Biodiversity Action Plan (NBAP). Upon public pressure, the revised draft was made available on the MoEF's website on August 31, 2007, asking for comments from all stakeholders. The revised draft incorporated very little of the NBSAP's final technical report. In November 2008 the Union Cabinet gave its approval to the NBAP. In attempting to advocate for the plan's acceptance or to make the findings of the process public, the TPCG approached the UNDP to advocate on its behalf. The UNDP could only take the MoEF on through soft dialogues that were unsuccessful. The TPCG also attempted dialogue with the ministry through the Prime Minister's Office, the National Advisory Council and members of the Planning Commission. The *Lok Sabha* (the House of the People) of the parliament also demanded reasons for the ministry's rejection of the report. In spite of engaging these different pressure points, the MoEF's own report was published without further consultation with the TPCG.

The coalition in the face of this stalemate petered out and then fell apart. One of the major weaknesses of the coalition was its short-term perspective, which blinded it to important aspects that should have been taken into account, including the specific framing of issues, the degree of criticism made of state processes (less would have been better), more involvement of academics in the technical and core policy group, and the garnering of more involvement from politically influential groups, like industry. For instance one recurrent theme found through the process of opposition by the ministry was the idea of the value of scientific knowledge over local or indigenous knowledge. The NBSAP coalition placed paramount importance on the experience and language of locals as representatives of nature. This was often in conflict with the framing of policy documents that officials claimed should reflect the science and tenor of official government documents. The positions taken by the ministry, on the other hand, seem indicative of a non-transparent and non-accountable policy process, where criticism of its current development paradigm was unacceptable. It is clear, however, that advocacy was undertaken within parliamentary means with no call for the masses to mobilize to demand that their plan be accepted. In addition there was a fundamental weakness with the process, in the sense that they did not bring influential actors, like industry, on board, who could have pushed their agenda at the state level. Because of their inherent suspicion of market forces, they came into conflict with politicians and bureaucrats who have an underlying emphasis on economic growth and are reluctant to alienate the interests of other influential actors, like industry. By being unable to push through the pro-people's livelihood agenda, the TPCG opened itself up to questioning by stakeholders as to the legitimacy of their role in the process.

The NBSAP's formulation process was mainly centered at the community and grassroots level, aiming to include marginalized voices into political processes; however, they failed to build up an active constituency that would garner support and advocate against the government once the policy was rejected at a central level. This could be because of several reasons:

1. Many considered that, as a technical issue, biodiversity could not be adequately tackled by the TPCG, which mainly consisted of social scientists and activists with a smattering of natural scientists. It was pointed out that the very thing the TPCG advocated for – the voice of the common people – diluted an issue, which the ministry and others considered a "technical" issue. Critics pointed out that the NBSAP members had no concept of data management. Many people contributed at all levels including state plans, and people often contributed unpublished material, seeing it as an opportunity to publish it with no system of analysis. Thus, it "remained a kind of word exercise rather than an analytical exercise" (member of BSAP, personal communication, June 25, 2009).
2. The network of the Technical and Policy Core Group are a group with loose affiliations that coordinated action for the first time. Though they did have a core belief in community rights and participation, they had many differences in the form and issues the NBSAP integrated. As it was a mix of professionals from various sectors, time constraints, differences in priorities and varied dependence on the state kept them from forming a strong advocacy group once the state became less responsive.

3. A TPCG member noted that there was an inherent unfamiliarity with concepts of participation and biodiversity. This type of process, the first of its kind in India, needs an informed and reactive citizenry. In one interview with a TPCG member, it was pointed out that "Indian citizens are not used to calls for participation. Second, it is on biodiversity, and people still have weird notions about it, it is definitely seen as a specialists' sector" (Apte 2005, 113).

The coalition advocating with the state on the NBSAP had a dynamic vision of rights. The NBSAP wanted constituents to articulate their own rights and wanted the policy document to reflect the diversity of voices and interpretations. It attempted a highly participatory formulation process that attempted to reach out to "a large number of village level organizations and movements, NGOs, academicians and scientists, government officers from various line agencies, the private sector, the armed forces, politicians and others who have a stake in biodiversity" (Kalpavriksh n.d.). In addition to more formal methods of research like questionnaires, thematic working groups and academic papers, it also held public hearings, biodiversity festivals, school involvement, cultural programs, and cycle and bullock cart rallies. The technical and core policy group functioned as a formal coalition using a flexible methodology, while also attempting to interpret the formal modes of participation within the policy process.

The NBSAP broke away from more traditional modes of representational participation. They shared a common belief in protecting biodiversity and the rights of Adivasi and rural communities and did not follow the more *acceptable* route of policy formulation, by which they could have represented the communities they advocated for, giving the communities themselves minimal representation.

In the case of the NBSAP, the document in the process of formulation was discussed at the state and local level to various degrees, in different forms and context. Local-level participants attempted to understand and interpret integrated biodiversity in local terms. In the case of the final NBSAP released by the ministry, it was put up for deliberation on the ministry website "inviting comments of all stakeholders. The Draft also circulated to concerned Central Govt. Ministries/ Departments for comments" (MoEF 2011, COP10). Thus, the states interpretation of participation was much narrower that what had been envisioned by the TPCG.

The NBSAP was labeled "unscientific" (Kohli 2009) and "too diluted" (MoEF official, personal communication, April 19, 2010). In a written response to a parliamentary question on the rejection of the NBSAP, the MoEF stated that the plan contained inaccuracies that would embarrass India internationally, but it did not attempt to discuss these with the TPCG. The draft NBSAP was first reduced to a "technical report" (2004) and then further demoted to a "consultancy report." In 2009, MoEF authored its own version of the National Biodiversity Action Plan (NBAP), written by MoEF officials. The government mentioned that it was decided that the NBSAP could be finalized only after the National Environment Policy (NEP) (which at that time was under preparation) was finalized and approved so that it would be in conformity with NEP. After approval of the NEP by the cabinet in May 2006, preparation of the NBAP was taken up, using the final technical report of the NBSAP project as an input. The draft NBAP was put on MoEF's website in September 2007.

Some critical additions and emphasis appeared, including valuation of goods and services provided by biodiversity and the use of economic instruments in making decisions that impact biodiversity. In one scathing review, a member of Kalpavriksh notes:

> Very disturbingly, the NBAP in several places also justifies the use of bio-technology, which promotes monoculture and stands in complete contradiction to the variety of life forms the term 'biodiversity' encompasses within itself. . . . In classical parlance the word bio-prospecting is embedded in the language of access for the purpose of trade, which is what the MoEF has in store for India's biodiversity.
>
> (Kohli 2009, last para.)

Others criticizing the dilution of the role of community in the final MoEF policy pointed out, "Community leaders are now 'data providers' while scientists and government are 'validators'. Except for ascertaining scientific names, 'experts' have little association with the natural knowledge they are documenting" (P.V. Satheesh of Deccan Development Society, TPCG; quoted in Acharya 2007, para. 8).

This criticism is well placed as the MoEF as recently as 2011 in Japan[8] highlighted commissioning a "national study on economic valuation of biodiversity" and pointed out that future targets are enshrined in the language of ecosystem services and the integration of economic value of biodiversity. At the same time, the MOEF continues to refer to the failed NBSAP process "as one of the most participatory environmental planning processes, facilitated by MoEF through unique consortium arrangement, in an attempt to move away from general trend of centralized planning" (MoEF 2011). They detail its participative elements and unconventional methodologies, like biodiversity fairs and organized rallies. It mentions that it was "[u]nique process applauded even at international level" (MoEF 2011).

Conclusion

The failure in deliberative democracy even in such a participatory process is easier to understand when one sees it as a symptom of a wider trend in environment policy processes in India, which gives primacy to economic growth rather than environmental conservation and livelihoods. The broad purpose of the NBSAP remained close to the CBD in spirit, namely a broad implementable plan focusing on conservation of biodiversity resources, livelihood security and a participatory process leading to sustainable utilization of these resources and deriving decision-making on both access and utilization of the same. These encapsulated different agendas at different levels of governance and of the different actors involved. The most contentious of these norms that were filtered down, advocated and implemented remained the participatory planning process, which was advocated for by NGOs who had seen the NBSAP as a window of opportunity to prove that involving people brings its own benefits to environment policy processes. In a scenario, where the government retains a heavy-handed approach to environment decision-making, they were successful in creating local implementation strategies, capacity building and an increased awareness on the issues surrounding biodiversity. They were also successful in highlighting the weakness of government strategies in this area.

At the local level, one of the main highlights was in Karnataka, where the Centre of Ecological Sciences was able to incorporate many of the recommendations of the Biodiversity Strategy Action Plan (BSAP) into the Karnataka State of the Environment Report. A coordinator of the process felt that in this instance the BSAP had not only managed to create a productive synergy with an official process but it also affected its implementation as people had a greater sense of awareness of the issues through the *uttar kannada* and came forward with well thought out and concrete recommendations. Others felt that Karnataka was a success story because it proved that if you present recommendations that can work in a concise and *technical* format, the government will not be averse to implementing them.

In the wake of the failure to officiate the NBSAP, it was also found that political "buy in" is a necessity. Though the TPCG focused on participation at a grassroots level, NGOs and activists, it was found that very little effort had been made to engage with politically influential actors. This became a major weakness in the process of advocating for the NBSAP at a national level and highlighted the need to target priority groups along with large-scale participation. Thus, this case study throws up the question of whether true deliberative democracy is achieved just by virtue of the sheer numbers participating. One then wonders if the deliberative element of participative initiatives is often not capitalized upon. Though the structured forums of different levels at a national, state and sub-state level are harnessed by particular actors and equal weight given to both technical and field experience, can the nature of deliberation be sustained through such large-scale participation without the role of influential representatives?

This case study discusses broad participatory approaches that aim at giving citizens direct access and influence in political decision. The NBSAP was one of the more participatory environmental planning processes, facilitated by MoEF through a unique consortium arrangement, in an attempt to move away from the general pattern of centralized planning. However, its failure indicates that the role of political representatives at certain key junctures of the political process is necessary. In addition, the tensions between direct and indirect democracy are evident; involving representatives like politicians runs the risk of turning a potentially rewarding experiment into a political performance that reflects certain structural issues. This case study suggests that the government's strategy has changed in the last sixty years. Development priorities have shifted to a narrative of ecological modernization. Policies and plans, because of the influence of international regimes, have also shifted to ideas of management rather than resource claims or upholding of rights. Though the state talks of "participation" and "decentralized planning," they are often loaded terms, playing out differently when the state directly engages with what this narrative underlines. As a result, while it is necessary to engage with the process of fighting for community rights to land and resources, there are several obstacles that one has to overcome. This case study shows the lines of conflict over participation. It underlines the tension between dominant narratives in the emerging space of biodiversity, where rights-based groups with alternative visions of development have launched their struggles. These struggles have conformed to the deliberative space set out by the state, both in their engagement with rights and in their advocacy against a mainstream narrative. However, one sees that within this space, if groups must be successful, they must engage more deliberately with market-based discourses or build enough social capital. This will pressurize the state to accede to demands from beyond the boundaries of the state and force it to conform to their vision.

Notes

1 This is supported by Tandon and Mohanty (2002, 52), who point out that "the national and the provincial level the bureaucracy is vested with the power to formulate the policies which are expected to bring benefits to the common people. This process which does not involve the opinion of the people in decision-making fails in many cases to take cognisance of their needs."

2 The BSP is a USAID- funded consortium of the World Wildlife Fund.

3 Nature conservation policies are often based on the assumption that poor communities lack an appreciation for nature and pose a threat to biodiversity. Despite evidence that casts doubt on the link between poverty and the loss of biological diversity (Duraiappah 1998; Michaelidou and Decker 2003), the assumption that poor people are an impediment to biodiversity conservation has persisted in the international conservation arena.

4 Article 4.2, United Nations Declaration on the Right to Development (DRD), adopted by UN General Assembly resolution 411128 of December 4, 1986.

5 Interview with Madhu Sarin, human rights activist and member of the NBSAP technical and core policy group (May 5, 2010).

6 A. Sen, director, Planning Commission, "Ensuring Food Security in a Changing Climate," organized by Gene Campaign, lecture conducted by Gene Campaign and Action Aid (New Delhi, April 24, 2010).

7 MoEF, Guidelines for selection of non-officials to various statutory, judicial, and other committees and other bodies in MOEF, 20011/3/2003 – GC, ministerial circular, 2004.

8 MoEF "Biodiversity NBSAP: India's Experiences," presentation by India: Global Launch of United Nations Decade on Biodiversity, Kanazawa, Japan, December 17, 2011, http://isp.unu.edu/news/2011/united-nations-biodiversity-decade-launched-in-kanazawa.html (accessed June 12, 2012).

4 Deliberating on the Forest Rights Act

Forests in the developing world are highly contentious spaces, the object of competing resource use claims and ideologies. Historically in India they are government controlled, and very little scrutiny is allowed on competing resource-use claims, so much so that even sixty-two years after independence, in reply to the query "What is a forest?" filed through the Right to Information Act (2006), the Ministry of Environment and Forests replied that the "definition of forests is under active consideration and has yet to be finalized" (*Daily News and Analysis* 2011, para. 3). In this context, forest-dependent livelihood communities, especially in complex bio-diverse habitats, are pitted against both public conservation efforts and powerful economic and political interests. Frequently the same communities are blamed or expelled from the resources that they have been dependent on for generations.

Poffenberger et al. (1996) has found that there is a significant overlap between the location of forests and the presence of poor communities, many of whom are tribal people. Around 275 million people in India are estimated to be dependent on forests for the majority of their income. These communities, who are termed forest dwellers, are groups that have the highest incidence of poverty in India (World Bank 2006). Mehta and Shah (2003) point out that 84 percent of India's tribal ethnic minorities are engaged in forest-based economies and that they are the ones who experience the greatest poverty in society.

Inclusion of marginalized groups and minorities is one of the most conflictual aspects of democratic deliberations. As we saw in the previous chapter, in the NBSAP process the Technical and Policy Core Group resurrected the rhetoric of participation, acting as cultural brokers, translating the norms of biodiversity into grassroots rights-based activism. The Scheduled Tribes and Other Traditional Forest Dwellers (Recognition of Forest Rights) Act of 2006, popularly known as the Forest Rights Act (FRA), brought the concerns of the marginalized into deliberative spaces within the democratic political system from the bottom up. In this chapter, I analyze the processes that led to this law and show how grassroots organizations representing traditionally marginalized voices advocated both within and outside the spaces of deliberation provided by a representative democracy to push right-based claims into a national legislation. This chapter also traces the ensuing conflicts of narratives, rooted in historically based environmental ideologies that surround such claims.

Theorists have argued that structural and cultural changes in industrial or post-industrial societies, as well as in less-developed and developing countries, are generating new types of expressions of defiance and mobilizations for change, often referred to as new social movements. Mobilization has been emphasized as a key requirement for political participation (Rosenstone and Hansen 1993; Verba et al. 1995) and underlined as one of the central processes for both social movements and other political intermediation organizations to promote collective action (McCarthy and Zald 1977; Snow and Benford 1988). Movements are characterized by their inclusiveness and limited formal organization, the articulation of a conflict with hegemonic and mainstream values, the articulation of rights and the sustained nature of these efforts (McAdam et al. 2001). This broad definition allows for the inclusion of different profiles of actions and actors with different strategies who work together towards long-term goals in both social and political domains and often develop formal organizations. Its success depends on several factors: the first is the way issues are framed; the second are structures of mobilization, which include formal and informal movement organizations and networks; the third depends on the windows of political opportunity available (McAdam et al. 1996). In order to capitalize on political opportunity and establish new social structures, movements must develop allies with elite groups who retain access to the political system, in order to minimize or escape state repression (Tarrow 1989).

This study contrasts collective action through two axes of participation – vertical and horizontal. Vertical spaces are those created by the state and which *invite* citizens to participate. Cornwall and Coelho maintain that "the institutions of the participatory sphere are framed by those who create them, and infused with power relations and cultures of interaction carried into them from other spaces" (2005, 11). These are "invited spaces" (Cornwall 2002, 17) in which communities have the potential to engage with the state constructively and would include Hendriks's (2006a, 486) "micro deliberative structures" and Fung and Wright's (2003, 5) "empowered deliberative democratic structures." However, as we saw in the previous chapter with the NBSAP, poor design of these spaces, a lack of will on the part of political elites and the relative lack of power of key social actors mean that, in practice, they can turn into meaningless processes.

In contrast, horizontal forms of participation are emergent processes in state-society relations from below – linkages forged between mobilized citizens and communities at the local, national and global levels. These horizontal spaces of participation, which might also be called "self-created" or "invented" spaces, are where citizens themselves define their modes of engagement with the state and with other interest groups and resort to different forms of collective action. These "invented spaces" created from below by those outside the state include spaces created from popular mobilization. Holston and Appadurai (1999) describe the emergence of a rights-based citizenship among the urban poor, marginalized by neoliberal governance and mobilized through social movements, which looks to transform social relations from the ground up. These linkages are not necessarily stable, nor do they represent a fixed notion of citizen identity on the part of those who participate. The ways in which mobilization, collective action and social movements manifest themselves in these spaces of state-society relations are a key to understanding the processes of collective identity formation as citizens attempt to exercise both their individual and their collective rights.

Individuals who participate in movements singularly adopt collective identities; they also become part of the political culture and public discourse (Mueller 1994, 256). Social movement communities consist of groups and individuals engaged in "ideologically structured action" (Zald 2000, 3). Within the environmental movement, there are several significant narratives, defining distinct movement sectors or "wings" which diverge in terms of their definition of problems, strategies and methods of organization (Brulle 2010, 396). As framing theory and "ideologically structured action" show (see Benford and Hunt 1992; Benford and Snow 2000; Zald 2000, 3), ideas and discursive frames shape a number of characteristics of movements. This case study makes an important contribution to highlighting conflicting discursive frames[1] around which mobilization occurs. It traces the mobilization of communities, who respond collectively to build strategies from the top-down to make space for rights-based claims within the deliberative space, struggling against the traditional command-and-control discourses of the state and exclusionary politics.

Mobilizations around rights of resource use and contestation in formulation of forest policies and implementation of the same have straddled both civil society and the state. In the case of the Forests Rights Act, actors like NGOs, grassroots campaigns and community-based organizations have been part of the process from inception to implementation and put the agenda of forests rights on the table. There are contradicting claims on who is responsible for the success of the act, with grassroots campaigns and Adivasi movements claiming that the success lay in the creation of a public demand for forest rights. The media has repeatedly underlined the wildlife lobby's claim that the act was more the handiwork of the forest mafia, who had struggled to exclude national parks and sanctuaries (particularly tiger reserves) from the purview of the act. Others have claimed that institutional support from the state allowed the act to go through.

In all its forms and conflicting demands, there is however no doubt as to civil society's mobilization around conflicting agendas. Over one hundred organizations and individuals submitted memoranda to the Joint Parliamentary Committee on the Bill. The drafting of the bill was undertaken under the purview of the state, involving the Ministry of Environment and Forests (MoEF) and the Ministry of Tribal Affairs (MoTA), which constituted a Joint Committee in April 2010 to review the implementation of the FRA, along with independent consultants who had been part of the process of formulating the policy. Hundreds engaged with policymakers during the rule's formulation. Still later, members of civil society filed six petitions in the High Courts and Supreme Court, calling for the act's annulment.

This struggle for democratic control over forest resources emerged from structural issues: insecurity of land tenure and access rights, lack of recognition of community conservation initiatives in forest management, lack of recognition of traditional governance and resource ownership in tribal areas, and threats to community lands and forests from development projects. The Planning Commission of India had highlighted the importance of resolving these issues through protective legislation to deal with the growing discontent, unrest and extremism in tribal and forest areas (Planning Commission 2008). This case study establishes three key arguments: one is that organized mobilization was crucial in highlighting rights-based claims of traditional forests dwellers. Second, mobilization coupled with opportunities in the political space reshaped participatory policy-making in India. Mass movements attempted to legitimize alternative approaches to conservation, infusing rights-based, people-centered development rhetoric into a sector that had been dominated by the fines and fences approach. Like in the case of the NBSAP, it provided a discursive opening for groups of rights-oriented actors, shifting the axis of domestic rhetoric from conservation towards a rights-oriented approach, even in the face of great conflict. Third, it used deliberative openings within state institutions as a site for political activism aimed at introducing legislative and institutional reforms.

Narrative background

India's forests and protected areas, which cover approximately 20 percent of the country's land (MoEF 2009, 2), are sites for intense conflict between different constructions of what a forest is and whom it should benefit. On the one hand, the state views forests as an important source of revenue and underlines the need for their scientific management. On the other hand are the forest-dependent communities, for whom forests hold a central role in their lifestyle and cosmology and who are opposed to state control. The third group is the conservationists who desire minimal human presence and underline the need to preserve wilderness and protect the wildlife, which have their own intrinsic value. All three groups have their roots in different influences on the Indian forests debate. In this section, I trace the different influences on Indian forest policy that strengthened the idea of a historical marginalization that became the key frame around which mobilization of the rights groups occurred. I analyze the occupation and de-occupation of narratives that weave through the policies that surround forests in India and the subversion and resistance to these hegemonic frames by the marginalized.

The state and scientific management of forests: colonial forest policy

The early days of British rule were characterized by a total indifference to the needs of forest conservancy until in the middle of the nineteenth century, when there was a "fierce onslaught on India's forest" (Smythies 1925, 6). It was only in 1856 that a definite forest policy was laid down, owing to difficulties in sourcing timber (Smythies 1925). The imperial forest department was formed in 1864, with the help of experts from Germany, who were leading Europe in forest management. The first attempt at asserting state control over forests was through the Indian Forest Act of 1865. This law allowed the state to appropriate any land covered with trees or brushwood as state forest and to bring the management of the same under the purview of the state. Historically, this has been seen as the first attempt to exert state control on the collection of forest produce by the forest dwellers, although private forests, belonging to the princely states, continued to remain outside state control (Kulkarni 1987). In 1856, Dietrich Brandis, a German botanist and forester, was appointed first inspector general of forests in India. It was Brandis who first introduced practical scientific forestry in India with an objective "to measure the annual growth of tree stands and to evaluate how much timber could be extracted annually without compromising the future productivity of the forest" (Stebbing 1923, 42–43). From the British perspective, the creation of a separate forestry service was a waste of money as the general imperial policy underlined that the administrative machinery in India had to be self-supporting. This created enormous tension with the imperial Revenue Department, and therefore foresters had to prove that forestry was an economically viable enterprise, which made the case for rationally or scientifically managed forests stronger (Guha 1983).

The Indian Forest Act of 1865 was passed after prolonged debate in the colonial bureaucracy. These debates echoed contemporary debates in India regarding state control versus community rights. Gadgil and Guha (1996) categorized the debate into three different strains of arguments. The first strain they term "annexationist," meaning that the state desired total state control over all forest areas. The second "pragmatic" strain argued in favor of state management of ecologically sensitive and strategically valuable forests, allowing other areas to remain under communal systems of management. And the third, termed the "populist" strain, completely rejected state intervention, reinstating sovereign rights of tribal people and peasants over forest. The colonial bureaucracy was divided between these positions, and, after several negotiations, the state's supreme rights over forests were established. This paved the way for the conception of rights, specifically those of a customary nature, to be framed as "privileges" rather than "rights."

In 1878, the British introduced the more comprehensive Forest Policy Act, which was largely aimed at harnessing forest goods for economic maximization. This policy, backed by the scientific community, divided forests into reserved forests (those entirely under government control), protected forests (where forest dwellers could enter to collect food and fodder but under government discretion) and unclassed state forests (where the government gave permission to forest communities to collect forest products for their household needs) (Gadgil and Guha 1996). Forest-dwelling communities were required to record their claims over land and forest produce in the proposed areas. Certain activities like trespassing or pasturing of cattle were prohibited, and timber duties were imposed on all forests, including private ones. Certain activities were declared as forest offences, and imprisonment and fines were also prescribed. This was then followed by a National Forest Policy in 1894 and then the Forest Act of 1927, which increased state control over the forests, removing them from the complex patchwork of community control, and placed more stringent regulations on people's rights by converting forests into government property for scientific management.

Post-colonial forest policies

Ramachandra Guha (1983) has argued that, before 1947, forests served the strategic interests of British imperialism and that, after independence, they served the needs of the mercantile and industrial bourgeoisie. Even as the Indian government attempted to redirect state policy and orient it towards welfare, the emphasis of forest management regimes continued to be on commercial timber exploitation and the exclusion of local people (Kant and Cooke 1999). The use of forestry in socio-economic and rural development continued to be emphasized in the five-year plans of the government. In the first two five-year plans (1951–56 and 1956–61), there was an emphasis on the regeneration and consolidation of forests that had been exploited but primarily in order to meet the fuel and raw material needs of industry (Planning Commission 1951, chap. 21). Maintaining the orientation of 1894 policy, forest management emphasized protecting forests in order to fulfill defense, railway and industrial needs. This was mirrored in the 1952 National Forest Policy that legitimized the idea of sustained yield by emphasizing that forests would be managed to meet the "paramount needs" of the nation (Ministry of Food and Agriculture 1952, appendix III, sec. 3). In the Third Five-Year Plan and the Fourth Five-Year Plan (1961–66 and 1969–74), there was a change from a focus on purely biological and scientific management of forests to the idea of increasing production from existing forests in India and also creating human-made forests to meet the raw material demands of industry. Special emphasis was laid on creating plantations and training foresters as resource managers. This is articulated in the plans' proposals to replace trees with only fuel value with "valuable planted forests" (Planning Commission 1961, chap. 12) and "make arrangements to provide for environmental expertise" (Planning Commission 1969, chap. 2). These plans "put special emphasis on measures which will help meet the long term requirements of the country and ensure more economic and efficient utilisation of the available forest products" (Planning Commission 1961, chap. 22).

In the Fifth Five-Year Plan (1974–79), there was a shift from production orientation to conservation with more stress laid on management of national parks, although extensive human-made forests were created in order to meet the demands of forest-based industries. In 1972, the Wild Life (Protection) Act was passed, creating the system of national parks and wildlife sanctuaries that we know as "protected areas" today. This act, with its exclusionist underpinnings, created many conflicts. Specifically it used the same system of settlement of rights that was present in the 1927 Indian Forest Act but imposed much more strict restrictions on people's use of forest resources (Wild Life Protection Act 1972, chap. 4, sec. 24). In national parks, for instance, all rights of forest-dependent communities were revoked. The act also allowed reserved forests to be converted into sanctuaries without any process of recognition or settlement of rights (Gopalakrishnan 2010).

The space for social and community forestry finally emerged in the 1980s where programs for social and community forestry were expanded, including large programs that recognized the central role of humans in the forests. In 1980, India passed the Forest Conservation Act that barred any de-reservation of forests, or use of forestland for "non-forest purposes" (MoEF 1980, sec. 1.4), except with the permission of the central government. The assumption was that state governments were exploiting forests for financial gains.[2] The state governments, however, blamed people's encroachment as the primary source of forest destruction. Corruption in the forest department and pressures on the central governments to grant waivers weakened the aims of this act. States protested, as they believed that the central government was placing unnecessary hurdles in the way of development projects. For forest-dependent communities, the prohibition of non-forestry activity on all forestlands and control of shifting cultivation and encroachments made their position even more tenuous. As a result, many forest-protection initiatives by local communities emerged in response to growing scarcities and threats of exploitation by outside groups. These community initiatives were a manifestation of conflict between the formal and informal institutions of forest management (Prasad and Kant 2003). In 1988, within the framework of sustainable development, the new National Forest Policy made a significant shift in forest policy by underlining the need to consider local community interests for utilization of forest resources. It also called for increased participation of local communities in the protection and regeneration of forests. In particular, the policy asked to find a balance between the dependence of communities on forests and the management of forests (MoEF 1988).

Reflective of this new era of increased participation, the government introduced the joint control of forests between the state and the people (joint forest management [JFM]) supplementing the National Forest Policy of 1988. Though some saw JFM as a new policy culture with the state finally embracing the notion of human centrality to forest ecosystems in India, others considered it nothing more than sharecropping arrangements between the state and local villagers. As Sarin (1996) points out, the structures of JFM, in both content and promise, only reiterated the inherent unequal relationship between powerful state bureaucracies and the forest-dependent communities. The forest departments continued to wield ultimate power over communities, as they retained the right to unilaterally cancel JFM agreements if the latter was perceived as violating any given condition. In addition, because of its limited mandate (afforestation), JFM failed to address the livelihood needs of farmers and also displaced marginalized farmers who cultivated degraded land that was appropriated by the state to be afforested (Sundar et al. 2001, 183–87). Within a decade, the novelty of the JFM had worn off, and more conflicts emerged around the competing claims to the forest manifested in two other process – the Godavarman Thirumpulpad or Forest Case and the mobilization around the enactment of what was eventually passed as the Scheduled Tribes and Other Forest Dwellers (Recognition of Rights) Act of 2006.

We thus find that attempting to strike the balance between efficiency and welfare became the cornerstone of post-colonial forest policy. Efficiency directed policy towards the scientific utilization of forests, while welfare aimed at the distribution of benefits from the use of these resources. In the efficiency paradigm, the focus was on trees while the people were marginalized, whereas, in the welfare paradigm, the focus was on people and it was assumed that the trees would follow.

Conservation politics

The issue of wildlife conservation began to be a priority from the Second Five-Year Plan (1956–61). Following the tradition of scientific management, many wildlife sanctuaries and parks were created, while others were upgraded. The idea of wildlife conservation in India has often been challenged because of the resources set aside which could be diverted to alleviating poverty or other development concerns. However, article 48 of the Constitution of India specifies that "the state shall endeavor to protect and improve the environment and to safeguard the forests and wildlife of the country," and article 51-A states that "it shall be the duty of every citizen of India to protect and improve the natural environment including forests, lakes, rivers, and wildlife and to have compassion for living creatures" (India Const. 1949).

In 1952, the Indian Wildlife Board was set up to centralize all the rules and regulations relating to wildlife conservation in India, which until then differed from state to state. In 1956, this board accorded all existing game parks the status of a sanctuary or a national park. In the 1970s, growing environmental concerns led to two new central laws: the Wildlife (Protection) Act of 1972 (WLPA), which included the declaration of protected areas (PAs), and the Forest Conservation Act of 1980 (FCA). In addition, the Government of India initiated a number of other programs for the protection and revival of populations of highly endangered large animals such as the rhino, elephant and tiger, the last of which was achieved through the launch of Project Tiger in 1973.

In India, there is no particular definition for a protected area; any area that is considered by the central government or state government to be important for conservation is designated a status under the WLPA and is then legally considered a protected area. This ambiguity is compounded by the fact that the state has taken over huge areas of customary tribal lands as state "forests" or "wastelands," and extended all its laws to them. Instead of withholding or adapting the present laws to accommodate the Adivasis' customary tenures and governance systems (Sarin 2010, 111).

Initially, the WLPA made provisions for two main types of protected areas – National Parks and Wildlife Sanctuaries (WLPA 1972, sec. 35). An amendment to the Act in 2002 included two more categories – Conservation Reserves[3] (WLPA 2002, art. 36A) and Community Reserves[4] (WLPA 2002, art. 36C). A further amendment in 2006 added another category called the Tiger Reserve (WLPA 2006, art. 38 V [4]). The WLPA prohibited people from living within the boundaries of these newly categorized national parks and sanctuaries. This lead to forced relocation of people and villages that were in national parks all over the country, as traditionally both rural villages and tribal communities lived in these areas. Unlike in the United States, which developed the national park model, India did not have open spaces with no or few inhabitants (Lewis 2004, 8). Just as the conception of scientific management, within colonial structures, was also an export of a particular conception of nature and resource use, many have argued that the parks that ecologists and environmentalists support are simply US cultural exports to India and the rest of the world (Lewis 2004, 14). This is well articulated by Saberwal and Chhatre (2002, para. 11), who clarify the context within which conservation debates in India take place:

> Political in so far as an identifiable constituency has attempted to push through the idea that all human activities are inimical to the conservation of biodiversity. Such a relationship is clearly not axiomatic. And yet, even in the face of evidence pointing to the contrary, there is little attempt on part of the mainstream conservation lobby to engage with alternative models regarding the impacts of humans on the landscape.

According to Lewis, these identifiable constituencies are ecologists and environmentalists who have succumbed to the influence of US ecology and academia on the conservation politics of India. The large number of theories, training sessions, publications, funding, scientists and collaborative projects omnipresent in the country's most influential ecological institutes represent this (Lewis 2004, 14). The idea of scientific management pushed by the colonial and post-colonial structures of environmental governance have gone hand in hand with the exclusionist model of inviolate areas, pushed by conservation groups who underline the idea that coexistence is not possible between humans and wildlife. Growing availability of international funding for conservation projects also encourages governments to adopt a language more in line with international ideals of conservation, which are in stark contrast to the realities of land-use patterns on the ground. In response, these same conservationists who push the "inviolate areas" or "fortress conservation" models are seen by rights-based activists or developmentalists as the elite "encultured within an exclusivist conservation ideology" (Mawdsley et al. 2009, 56) continuing the tradition of hunters and rulers who were concerned about depleting wildlife populations and who had "no connection with the common masses, nor did they understand their needs, knowledge and practices" (Broome 2011, para. 46).

The creation of protected areas dates back to 1864, when Yellowstone National Park was established in the United States of America. Following the same model, the first modern protected area in India was set up in 1931, called Halley's National Park (today known as Corbett National Park, in Uttarakhand). Most national parks in India still do not meet the law's criteria of no human habitation and no use of forest resources. There are many different spectrums and points of view engaged in a struggle to bring these parks in conformity with the law. Pro-people groups argue that people are not "inherently harmful to ecosystems," arguing that "Indian ecosystems have evolved within the context of human use" and that the solution lies within "the reconstruction of the forest for meaningful exploitation of these resources" (Lewis 2004, 9). Environmentalists argue that even within communities that live in harmony with the ecosystem, there is often a breakdown of traditional ways of life through migration, population changes or changes in the livelihood patterns that had made living in an ecosystem sustainable. With a breakdown of values or change in population structure, people more and more compete with the animals for scarce resources and shrinking forests.

As of 2011, about 5 percent of India's territory is covered by protected areas. The demarcation of these areas (as shown in Table 4.1) has kept many eco-logically critical areas and threatened wildlife species from being wiped out by dam projects, mining and urban and agricultural expansion. However, people, some of them ancient tribal communities, also inhabit this area. Studies indicate that 3–4 million people live inside India's protected areas.[5] Most of these people belong to communities who have lived for generations in the areas before they were designated as protected areas. These communities are dependent on local resources for fuel, fodder, medicines, non-timber forest produce and so on. In many of these areas, collection of non-timber forest produce contributes to more than 50 percent of each household's annual cash earnings. Continuing with the colonial conceptions on rights, these subsistence or commercial activities are recorded as concessions or privileges but rarely as a right. Because of lack of documentation, many of these people are considered encroachers, their activities unrecorded and therefore considered illegal, even though they have lived in the area for generations. The WLPA decrees that a process of settlement of rights must be undertaken before any protected area is notified, with livelihoods and habitation rights that must either be allowed or be acquired by providing compensation or alternatives. However, due to unrecorded rights and badly kept land records, this process has remained incomplete in most protected areas in India (Broome 2011).

The historical frames of conflict tied to conservation in India ran along many lines of debate, including rural vs. urban, scientific management vs. traditional knowledge, and fortress conservation vs. community-based approaches. Ultimately, these were aspects of a much larger struggle, which was the struggle for control over natural resources. These struggles, which played out over many decades, came to a head with the mobilization around the Forest Rights Act.

The history of resistance

The colonial and later post-colonial Indian state formally appropriated land and enclosed forests from the late nineteenth century, giving the activities of millions of people dependent on the forests for livelihood the legal status of "encroachers." In India, it is estimated that 200 million of the people are partially or wholly dependent on forest resources for their livelihoods (Khare et al. 2000). These forest-dependent groups in India contain both tribal and non-tribal forest users. The "Scheduled Tribes" (i.e. those recognized and "scheduled" under the Constitution of India) include over 84 million people comprising 8.3 percent of the nation's total population (Government of India 2001) and around one quarter of the world's indigenous population. Shah and Guru (2004, 8) explain that the "incidence of poverty, reflected by head count ratio, is higher than the all-India estimates for the majority of forest based states. . . . The pattern is more or less the same during 1993–1994 and 1999–2000." In addition, the extent of forest "dependence" is difficult to stipulate. There are several reasons for this: first, almost all rural households use forests or forest produce to some extent, which complicates setting a threshold as to what qualifies as

Table 4.1 Categories in forest conservation (source: Broome 2011)

	National parks	Wildlife sanctuaries	Conservation reserves	Community reserves	Tiger reserves	Critical wildlife habitats
Protection category	Very strict protection, no human activity	Strict, but allowing some human activity	Strict for activities negatively impacting conservation objectives, but do not impact human use	Strict for activities negatively impacting conservation objectives, but do not impact human use	Very strict protection from all human activity	Very strict protection from all human activity
Management and governance	Chief wildlife warden, Forest Department	Chief wildlife warden, Forest Department	Conservation Management Committee (representatives from the Forest Department, one representative from each *panchayat*, one representative each from Department of Agriculture and Animal Husbandry)	Community Reserve Management Committee (five representatives from the village *panchayat*, one representative from the Forest Department)	Forest Department	Forest Department
Human settlements	Not allowed	Allowed	Allowed	Allowed	Interpreted by the Forest Department as not allowed in the core	Interpreted by the Forest Department as not allowed
Rights to forest resources	Not allowed	If cannot be settled and/or alternatives cannot be provided, then allowed	Allowed	Allowed	Interpreted by the Forest Department as not allowed	Interpreted by the Forest Department as not allowed

real dependence which will adequately capture the real income from forests; second, the seasonality in the use of forests produce keeps fluctuating across households and seasons; third, the forest-livelihood linkage, because of its historically criminalized undertones, is often conducted furtively (Angelsen and Wunder 2003).

The Fifth Schedule of the Indian Constitution protects the tribal populations through legal mechanisms of ownership over lands and resources in the areas earmarked as the scheduled areas. This is a constitutional safeguard that deals with "Provisions as to the Administration and Control of Scheduled Areas and Scheduled Tribes" (India Const. 1949, schedule V, art. 244[1]). It also prohibits or restricts the transfer of land by, or among, members of the Scheduled Tribes in such areas, and it regulates allotments of land. The 73rd Amendment Act (act no. 40) of 1996, which is known as the Panchayats (Extension to Scheduled Areas) Act, also came to be adopted in most of the states with Scheduled Tribes. The act clearly states the importance of the *gram sabha* (the decentralized unit of governance in the tribal areas) in the scheduled areas and right to self-rule and governance of the tribal people. It empowers the *gram sabhas* to have control over resources and the right to "customary law, social and religious practices and traditional management practices of community resources" (Devullu et al. 2004, 6). Over time, however, the *panchayat* slowly transformed from a system of local governance to one of a state-regulated representative democracy (Sarin 1993), transforming them from village councils focused on day-to-day decision-making to highly politicized units.

The historic marginalization of forest-dependent communities by the state through incremental laws and the powerful exclusionist conservation policy trajectories that followed has been described in the previous section. The Commissioner for Scheduled Castes and Scheduled Tribes, in his 29th Report, said: "The criminalisation of the entire communities in the tribal areas is the darkest blot on the liberal tradition of our country" (Government of India 1992). Their livelihood practices such as hunting-gathering and shifting cultivation were termed "unscientific" and a "social evil" (Sahu 2012, para. 2). They were, in many instances, forced to engage in agriculture or become bonded labor. Sahu (2010) pointed out that historically there were surges of evictions of tribal communities, which were evident in an account by anthropologist Verrier Elwin, who, for many years, researched the Baiga in central India. Sahu (2010) also points to many violent battles of resistance by tribal communities that were waged in India's Adivasi (tribal) heartland, sometimes in the form of setting fire to the forest they were displaced from (e.g. 1916 – Uttarakhand forest-fire [Guha 1989]) or the Bhumkal Rebellion (1910) in Chhattisgarh, where there was a violent uprising of tribal communities who responded to dispossession from their land. This was caused because of a unilateral decision by the state to create "reserved" forests, increase land rents and demand free labor, all of which was compounded by two crippling famines. Others like the Halba Rebellion were long-term struggles (1774–79) against the economic exploitation, through tax and foreign interference, of both the East India Company and the local Maratha forces. Another prominent rebellion was the Munda Uprising in Chotanagpur (1899–1900), against land alienation and forced indentured labor that created respected leaders like Birsa Munda. This led to the Chotanagpur Tenancy Act of 1908 that provided some recognition to their *khuntkatti rights*[6] and banned forced labor. Other important tribes involved in revolt in the nineteenth century were *Mizos* (1810), *Kols* (1795 and 1831), *Daflas* (1875), *Khasi* and *Garo* (1829), *Kacharis* (1839), *Santhals* (1853) and *Nagas* (1844 and 1879).

In response, legislation was enacted to protect indigenous people's lands both in colonial and in post-colonial eras. The 1908 "Chotanagpur Tenancy Act" in Bihar, the 1949 "Santhal Pargana Tenancy (Supplementary) Act," the 1959 "Andhra Pradesh Scheduled Areas Land Transfer Regulation" with the amendment of 1970, the 1960 "Tripura Land Revenue Regulation Act," the 1969 "Bihar Scheduled Areas Regulations," the 1970 "Assam Land and Revenue Act," and the 1975 "Kerala Scheduled Tribes (Restriction of Transfer of Lands and Restoration of Alienated Lands) Act" are all state laws aimed at protecting Adivasi land rights (Bijoy 2003).

After independence, under article 342 of the Constitution of India, 255 tribes in seventeen states were declared to be Scheduled Tribes with special protective, political and development safeguards. However, over the decades it became clear that the benefits of development were unevenly distributed. The costs of nation building and development were borne disproportionately by the tribal communities. In post-independent India, tribal movements can be classified into three categories: (a) struggles against economic exploitation, for instance militant movements in Bihar demanding freedom from the exploitation of moneylenders; (b) struggles against outsiders exploiting their natural resources as in the Chipko movement, a movement of peasants and forest dwellers in the Himalayas who campaigned against deforestation of the forests on which their livelihood and safety against floods depended; and (c) struggles of identity and separatist politics as in the movement for creation of a separate state, named Jharkhand. In recent years these struggles have been more pronounced against development projects like dams (Narmada Bachao Andolan), mining (anti-mining movement in Kashipur, Orissa) and special economic zones (Maharashtra, Bengal, among others). The movements rally around ideas like "Loha Nahi Anaj Chahiye" (We want grains not iron), "Jal, Jungle aur Jamin Hamara Hai" (Land, forest and water belong to us) and "Jan denge, Jamin Nahi Denge" (We will surrender our lives but not our land).

Historically, a culture of mobilization was present amongst the indigenous people of India, who are by no means a homogenous group. Centuries of targeted marginalization and pockets of resistance to hegemonic ideas of development and forced displacement manifested in a nationwide mobilization to advocate for the Forest Rights Act. In contrast, since the 1980s, the Naxalite-Maoist insurgency has taken up arms against India's government. They view it as a struggle against widespread poverty and oppression perpetuated by the policies of the elite, who have exploited their land and alienated them from their forests and livelihoods. The armed Naxalite movement is also a predominantly rural or indigenous struggle that has spread over the states of Chhattisgarh, West Bengal, Karnataka,

Orissa, Andhra Pradesh, Maharashtra, Jharkhand, Bihar and Uttar Pradesh. In 2006, Prime Minister Manmohan Singh called the Naxalites "[t]he single biggest internal security challenge ever faced by our country."[7] Thus, we see simultaneous manifestations of centuries of oppression and discontent – one a peaceful mobilization to carve out a legislative niche that would represent indigenous people's demands, and the other a violent mobilization in demand of those same rights.

Sites of environmental conflict

As seen in the previous section, conflicts over meanings and objectives over forest use and control of natural resources is historical. Contestation can range from daily problems of access to ideological underpinnings and tensions over development and conservation. Civil society responses to these encompass a wide spectrum of activities from advocacy, dialogue and deliberation to courts and armed responses. As Sundar (2009, 7) points out, "[T]he arenas where these conflicts have played out have varied – from the streets to the courts, to the streets back again." One of the sites for contestation, which spilled over to mass mobilization, was in the courts: the Godavarman case.

The Godavarman case originated as a dispute between an estate owner and the forest department in Tamil Nadu (Godavarman Thirumulpad vs. Union of India, WP [Civil] 202 of 1995) to protect a part of the Nilgiris Forest from deforestation by illegal timber felling.[8] Since then, over 2000 "interlocutory applications" (separate writs) have been filed, expanding the scope from ceasing illegal operations in particular forests into a reformation of the entire country's forests. The courts defined forests as not just forests as mentioned in government record but all areas that are forests in the dictionary meaning of the term irrespective of the nature of ownership and classification. By doing this many considered that the court had "gone far beyond its traditional role as the interpreter of law, and assumed the roles of policymaker, lawmaker and administrator" (Rosencranz and Lele 2008, 11). The court, supported by conservationists, in 2002 set up the Centrally Empowered Committee (CEC) and through a court-appointed amicus curia[9] (in this case Harish Salve) suggested that states were allowing encroachments despite the court's directives. The scrutiny of the Supreme Court spurred the MoEF to unilaterally issue a directive on May 3, 2002, to all chief secretaries, forest secretaries and principal chief conservators of forests of all states and union territories. It asked the states to summarily evict all encroachers of forests "not eligible for regularisation"[10] (post-1980) and to complete the process by September 30, 2002 – i.e. within five months. This directive was both impractical and in complete contradiction to MoEF's own earlier (1990) set of six circulars, which were detailed guidelines on encroachment.

These May 2002 MoEF circulars led to a series of ruthless evictions in various parts of the country. This sparked protests and hardening attitudes against the court and the state in tribal areas already under the influence of Naxalites, who highlighted the issue of encroachment (Rosencranz and Lele 2008). Since people with unrecorded rights had no legal recourse and were labeled "encroachers," the court's decision to evict them from forestland led to large-scale evictions, the last of which heightened in scale after 2002, when more than 300,000 families were evicted. According to the National Forum for Forest People and Forest Workers (NFFPFW) (NFFPFW 2008, 6): "A marauding Central Empowered Committee (CEC) constituted by the Supreme Court and staffed with forest officials and hardcore wild-lifers and conservationists added to the muddle. The CEC went on going around the country issuing eviction orders at will."

In response, questions were raised in parliament about the process of regularizing encroachments. The MoEF came under pressure from political leaders from tribal belts. The chairperson of the National Commission for the Scheduled Castes and Scheduled Tribes, Bizay Sonkar Shastri, wrote to the prime minister on September 6, soliciting his direct and immediate intervention (*Down to Earth* 2003). Civil society pressure was also mounting with prominent figures and organizations protesting against the government's actions, and the same organizations and individuals later formed the forefront of the mobilization around the FRA. In addition the CEC came under criticism for its composition with "three officials from MoEF and two NGO representatives with a pronounced inclination towards wildlife protection" (Thakuria et al. 2003, para. 16).

In the face of so much criticism, the MoEF issued a clarification on October 30, 2002, stating that there had been no change in the ministry's stand on pre-1980 encroachments that were eligible for regularization. The fact that "the concern for tribals came four months and 23 days after the May 3 order" kept civil society in an uproar (Thakuria et al. 2003, para. 18). This led to further softening of the MoEF's stand. A ministry official clarified in the Lok Sabha in 2005: "While examining the issue of settlement of disputed claims of tribals and forest dwellers on forest lands, and eviction of in-eligible encroachers from forest lands in pursuance of the Supreme Court order dated 23-11-2001, it was observed by the Central Government that the State /UT Governments were not able to distinguish between the encroachers, and the original tribals and other forest dwellers living on forest lands."[11]

The ambiguous nature of forests in India has come from a reluctance to define just what a forest is, and what, and for whom, it is necessary. These lacunas in definition allow powerful interests to be played out in the interests of how a forest should be used. We find in the Godavarman case different trajectories of intervention, including by the ministry, the court, the media, civil society and the Centrally Empowered Committee. They were in conflict with one another, supported by powerful lobbies of conservationists, tribal activists and the state, whose positions were rooted in distinct historical narratives.

Building collective action

Deliberative theory claims that collective decisions are legitimate to the extent that they emerge from dialogical and reason-guided processes of public discussion among citizens (Benhabib 1994; Dryzek 2000; Freeman 2000). Such public deliberation may take place in formal, highly structured settings established for just that purpose (Fishkin 1991), or it may unfold in informal, diffuse settings spread out across the countless associations of civil society (Habermas 1996). Like in the process for advocating for the National Biodiversity Strategy and Action Plan, the organizations involved in mobilization around the FRA used both structured and diffused settings. In addition, these organizations used a repertoire of social movement actions – actions that persuaded with both the weight of their numbers and their capacity for material damage (Della Porta and Diani 1999). Their actions included marches, rallies, petitions, letter writing and mobilizing voters. Like democratic political processes, such actions attempt to persuade elites that there is large public support for or against a particular policy.

Citizens figure in this account of democracy as co-participants in a process of reciprocal justification and persuasion. They seek, ideally, to converge toward a rationally motivated consensus. This deliberative view, with the slant of social activism, retains the "talk-centric" conception of democracy (Kymlicka 2002, 290), and it includes social conflict, strategic interaction, the mobilization of pressure, and other such factors that are centrally important in democratic politics (Shapiro 1999; Walzer 1999).

In Marcuse's (2005) assessment of the political significance of the social forum, "major transformative change" is equated with seizing (electorally or otherwise) state power at the national scale. If social forums (or social movements) do not aspire to this outcome, then there is no question of "transformative social change." Second, the political efficacy of movements depends on their capacity to produce a common platform. Third, there must be a targeted system of oppression, overturning the common objective that unites all issues and struggles (quoted in Conway 2005, 426). The idea of creating a common platform and locating the common system of oppression captures the organizational aspects of social activism and mobilization. Such activism incrementally builds a forum that is poised to launch into a movement, when the opportunity presents itself. Social movement theory often explains the transformation of politically quiescent organizations into social movement resources by factors external to the organization, the most influential approach being the concept of "political opportunity" (McAdam 1982; Tarrow 1994). Political opportunity attempts to capture the relationship between movement organizations and the state. It argues that organizations mobilize when they can capitalize on changes in the state and its relation to these changes. This perfectly captures the creation of the National Forum for Forest People and Forest Workers, which began as a forum to organize all unorganized sectors of workers that launched into the mobilization around the FRA when it perceived a political opportunity. The conception of civil society as spontaneous organizations – like the Campaign for Survival and Dignity (CSD), which was born as a direct reaction to state atrocities against the tribals – is captured by Habermas's (1996, 367) idea of civil society, which he sees as

> those more or less spontaneously emergent associations, organizations, and movements that, attuned to how societal problems resonate in the private life spheres, distil and transmit such reactions in amplified form to the public

sphere. The core of civil society comprises a network of associations that institutionalizes problem-solving discourses on questions of general interest inside the framework of organized public spheres.

Thus, we find that the forum becomes a flexible but organized space that can accommodate both long-term organization as well as more spontaneous movements, combining to thrust their issues to the center stage when there is political opportunity. In the case of the FRA, this opportunity presented itself at the critical juncture when Adivasi property rights were caught between the court and the ministry. The CEC recommended discontinuing all future regularization, even allowing for the possibility of "excessive use of force, un-provoked firing, and atrocities punishable under the SC/ST Atrocities Act" in the process of eviction (Sundar 2009, 6). Evictions took place across the country, involving the burning of Adivasi houses and standing crops, destruction by elephants, and so on (Sundar 2009). Public mobilization compelled the ministry in October 2002 to ask states to constitute joint committees of officials from the Revenue Department, Forest Department and Tribal Welfare Department to settle disputed cases. For one year, little action was taken. However, given that this was a critical issue in securing the Adivasi vote and because elections were due in 2004, land rights to the Adivasis were soon incorporated into every party's manifesto. On February 5, on the eve of the elections, the MoEF under the Bharatiya Janata Party (BJP) government gave orders de-reserving forest land in order to regularize tribal encroachments up to 1993. The Supreme Court blocked this order, as it violated forest laws (Venkatesan 2004; Sundar 2009).

Creation of the forum

In the 1990s, forest workers began to be organized, and a separate national forum, the National Centre for Labor, started. The NFFPFW initially situated itself within the larger labor movement. It was formed in September 1998 in a meeting organized in Ranchi in the state of Jharkhand. It was attended by 120 representatives of organizations working with forest workers from nine states (Uttar Pradesh, Jharkhand, Chhattisgarh, Uttarakhand, Madhya Pradesh, Bihar, Gujarat, Maharashtra and Karnataka) and by intellectuals such as Dr. B. K. Roy Burman and Dr. Ram Dayal Munda.

The formation of this forum was the culmination of a process that started in 1993–94. The discussions held prior to the establishment of the National Centre for Labour (NCL), a national federation of unorganized sector workers, stressed the need to make special efforts to organize certain sections of the unorganized sector workers: particularly fish workers, home-based workers and forest workers. The challenge these workers faced was that despite forming a majority of the total workforce, their economic contributions remained largely unrecognized. It was therefore felt that a two-pronged strategy had to be adopted: to build a wider organization of all unorganized sector workers and to organize them sectorally. The National Fishworkers Forum had already been formed. There was a similar need to give greater coherence to home-based workers and forest workers. This was probably the first time in India that attempts were made to organize a national forum of forest workers, which also tried to ensure the participation and representation of all forest-based communities.

The NFFPFW, a Marxist organization whose circulars greeted participants as "comrades," was understandably critical of global and national liberalization models being pushed by the Indian state. They saw their role as "resisting against the capitalist onslaught all over the world" (NFFPFW Conference 2009). The catchwords of the larger movement were framed in terms of a revolution as captured by Sadanand Menon (2009, para. 3), who writes: "There is a rousing music-video by documentary filmmaker K P Sashi which has become popular. . . . It compresses the plight of [Adivasis] into a six-minute montage, with the main slogan 'We won't give up our lands, we won't give up our forests, nor will we give up our struggle.'" Writings emerging from the movement posed a critique of the mainstream capitalist development model, which according to them had taken just a few decades to destroy the forests and rivers that indigenous people and their ancestors had preserved for centuries. The rhetoric of revolution was nurtured as the activities of the forum gathered momentum and the forum prepared to present its demands to the state. The NFFPFW also stated another reason as to why joining under a common banner became important. The neo-liberal globalization and economic structural adjustment policies of the 1990s, in their opinion, "posed a direct threat to forest resources" (Chowdhury 2009, 9). In their paper, NFFPFW attacks the initiative of Joint Forest Management (JFM) and Community Forest Management (CFM) projects, which, they state, "were started in several states in which the World Bank and other international financial institutions which are controlled by rich countries started making large investments. The main goal of these projects was to destroy and replace the traditional structures of forest-based communities in the name of protecting forest resources. It became crucial that such anti-people measures be opposed nationally and internationally" (Chowdhury 2009, 10).

In 1996, fifty representatives from eight states participated in the first general conference of the NFFPFW. This conference strove to unify local struggles into a national alliance. It was here that the idea that India needed a specific law to guarantee "democratic structures of governance in forest areas in order to ensure civil rights and labour rights to forest-based communities" was born (Chowdhury 2009, 6). Under the forest policy of 1988, the Government of India introduced the elements of decentralization and local participation in the administration of forest. Under the purview of the same policy, the secretary of the Ministry of Environment and Forest, Sh. Sankaran, issued six circulars with detailed guidelines to ensure the rights of the local community. However, these remained mainly on paper, never backed by strong, enforceable legislation. By the second meeting in Lucknow in 1997, the number of participants had burgeoned to seventy representatives from nine states and a number of intellectuals and activists. Amongst other resolutions, there was agreement that the term "forest worker" should be more clearly defined and people's organizations and movements and representative from all regions would be identified and involved. The National Forum also stated that it would work "towards developing the collective consciousness and build up a comprehensive ideology for Forest Right movements, respecting and protecting the diverse identities and cultural traditions so that the forest based working people can achieve their human and labour rights, within the framework of Indian Constitution, UNHRD and ILO conventions"(Chowdhury 2009, 9). A follow-up workshop in Ranchi in 1998 saw the participation of over 120 representatives and debated the terms of difference between Adivasi and forest workers. The forum's name – National Forest People and Forest Workers – emerged after intense debate. The name recognized the diversity of labour relationships between Adivasis[12] and other forms of forest workers, depending on the categories of labour they were engaged in. The main focus of the movement was "enactment of a legislation which recognized the sovereign control of those dependent on forests or forest resources" (Chowdhury 2009, 10). One of the blueprints for the proposed forest legislation was the Chotanagpur Tenancy Act of 1908, which prohibits the transfer of Adivasi land to non-Adivasis. The Ranchi workshop kick-started the process of organizing forest workers, and from it grew the formal identity of the NFFPFW.

In the meantime, the NFFPFW constituted a national committee (which was the formal body to take forward the conclusions and action plans), which decided to organize regional forms and workshops to strengthen partnerships from diverse regions within the country. It was also decided that forest areas, rather than state boundaries, would be the basis for organizing regional representation.[13] In 2001 and 2002, regional meetings honed the Charter of Demands (*Adhikar Patra*) of the NFFPFW and were finally passed in the National Convention, which was held in Nagpur in October 2002. Another result from regional forums was that "these meetings clearly brought to the fore the atrocities and exploitation by the Forest Department, locally and regionally, and exposed the realities of government schemes such as JFM and CFM. In many areas spontaneous mobilization started among forest workers opposing these schemes" (Chowdhury 2009, 12).

In 2002, in reaction to the evictions, another federation of tribal and forest community organizations from ten states in the Indian Union coalesced into the Campaign for Survival and Dignity (CSD) imbued with a new common cause. They came out with details of the evictions, the legal positions of indigenous people, how these were systematically violated and what should be done. Some of the members who had been working for CSD were also part of the National Front for Tribal Self Rule (NFTSF). The CSD remained open ended with members flowing both in and out. They depended on internal resources and contributions from forest dwellers and voluntary expertise. Full-time volunteers worked at both state and national levels.

Like the NFFPFW, larger decision-making was carried out through deliberation in forums where representatives of different states would gather. National-level strategies and umbrella positions were decided in these forums, which were replicated at different state and district levels that fed back through communication channels to the national level. Decision making in the coalition was carried out through processes of collective deliberation. National-level strategies were decided in meetings held once every few months in which representatives from various states would come together. Meetings were organized to chalk out programs and action, as well as to brief representatives who would attend the national CSD meetings. The coalition mainly framed the debates in terms of injustice to forest-dwelling communities (CSD 2003). These frames had already been invoked in previous grassroots mobilizations and were reiterated in the writing of anthropologists and the government. Taking strength from historical narratives of injustice and dispossession from lands, they countered all justification for environmental and conservation demands. Forest dwellers were portrayed as the protectors of the forests ravaged by *development* and liberalization activities. This was pitted against the "elite" conservationist narrative of those "who would like to build a fortress around our forests" (Lenin 2011, 2).

The CSD initially demanded implementation of the 1990 orders which were recommendations made by Dr. B.D. Sharma, a civil servant who was the Commissioner for Scheduled Tribes and Scheduled Castes (a constitutional authority), who reviewed the conditions prevailing in tribal areas and focused on the underlying causes of unrest – the lack of settlement of land and forest rights. Some of his recommendations, among others, were to give land titles to "encroached" settlers. The CSD gradually converted this into a demand for a new law (Sarin and Springate-Baginski 2010).

Two other developments took this movement to a critical juncture. First, in Kerala at the turn of the century, there were several starvation deaths amongst Adivasis. As a result, activists began large-scale mobilization on behalf of Kerala's Adivasis throughout the 1990s (Cheria et al. 1997, 65ff.). Their demands led on October 16, 2001, to an agreement between the state government and the leaders of the movement. The agreement stipulated that all landless Adivasis and those owning less than an acre were to receive up to five acres within one year. However, this agreement remained unfulfilled (Bijoy 2007). Second, in May 2002, as mentioned earlier, the MoEF issued a notification demanding that people vacate all forestland under "encroachment." This was per a Supreme Court ruling in 2000 that re-defined "forest land," and all land which was not marked as "agricultural land" in the revenue records, as forestland. This decision further strengthened the control of the Forest Department over vast tracts of land across the country. The MoEF interpreted the Supreme Court's orders as a direction to summarily evict "all illegal encroachment of forestlands in various States/Union Territories."[14]

This threatened the livelihoods of millions of people and led to direct confrontation between the state and its people. Protests and demonstrations began in various parts of the country; petitions were also filed against the government in courts. In response to the Supreme Court's CEC-empowered eviction drives, coordinated action was undertaken by the Adivasis and other forest communities in the many states. They started filing thousands of claims of ownership of the lands in the office of the respective district collectors. In other states, like Uttar Pradesh and Bihar, people retaliated by reclaiming land that was under the control of the Forest Department. This process of filing claims to their lands took the shape of a mass movement. The government came under tremendous pressure, and the deadline as per the notification kept on being extended. The government finally had to declare that all those who had been occupying forestlands before the notification of the Forest (Conservation) Act of 1980 would not be evicted. Thus, the demand for legislation, which would ensure the democratic rights of forest-dwelling communities, gained strength.

Amidst the clamor, the NFFPFW held its first national convention with more than 400 participants. In defiance of government evictions, the movement encouraged forest-dwelling communities to reclaim disputed forestland, which they believed historically belonged to indigenous communities. They stated that they wanted to "solve this problem democratically instead of depending on tumultuous, often unjust, judicial proceedings," especially since the Supreme Court had delegated all proceedings in this issue area to the Central Empowered Committee, which had no representation in the NFFPFW (NFFPFW 2008, 16).

Certain points were emphasized and pushed continuously by the NFFPFW: (a) establish community governance over forest resources, (b) resist commodification of forest resources (NFFPFW Conference 2009), (c) include the indigenous people's perception of forests as a resource into the legislation, and (d) bring community leaders from various forest movements into the forefront of the national struggle for achieving community control over forests. NFFPFW saw this as a particularly historic juncture for the forest movement (NFFPFW pamphlet).[15] The organizations of marches and of protests and the linking of Adivasi groups to the 2004 Mumbai World Social Forum, where 5000 members of the NFFPFW participated, allowed isolated people to locate themselves in a larger global struggle. The NFFPFW gave people the identity of being part of a larger labor force: "[T]hey not only met representatives of other communities engaged in struggle across India, but also comrades from other countries, particularly Latin America" (Chowdhury 2009, 16). This fits in with Ganz's (2002) idea that integrating local knowledge with facts and broader social theories helps communities identify their particular circumstances in a larger social and political context. It also enables members of grassroots groups to generate wholly new ways of thinking and plans of action – what sociologist Francesca Polletta (2002) calls the "innovatory" and "developmental" elements of democratic participation. Thus, we find that the forum becomes the center around which initial mobilization takes place, accommodating talk-centric deliberations, concretizing its position and ideological underpinnings, and building up concrete structures to sustain a long-term movement. It also becomes the space where the social movement in direct reaction to the CEC eviction drives is launched, as was the case for the Campaign for Survival and Dignity.

Political opportunity

Complex social issues – such as social exclusion, economic inequalities and community regeneration – cannot be addressed by hierarchical approaches to governance, while growing social differentiation has made the task of governing more difficult. Kooiman (1999, 2000) argues that government alone is incapable of determining social development. Strict forms of control and structures of hierarchical governance must give way to a more collaborative form, with a wide range of actors that cut across the public and several sectors, and operate across different levels of decision-making. We see in the background section that this shift was already coming about through the decades, with the Indian state's growing understanding that exclusionist policies in resource governance were failing. State policy was shifting to more collaborate forms of governance as seen with the JFM and community-management policies.

Kitschelt (1986, 59) argued that political opportunity structures function "as filters between the mobilization of the movement and its choice of strategies and its capacity to change the social environment." The crucial dimensions of these political opportunity structures are the openness or closedness of states to inputs from non-established actors and the strength or weakness of their capacities to deliver the effective implementation of policies once they are decided. This ties in to the discussion in the previous chapter on the Indian state itself as the "third actor" (Rudolph and Rudolph 1987). Although constrained by conflicting demands from different actor groups, the state has sufficient political authority and economic resources to carve out its own agendas. This means the Indian state also selectively chooses agendas to give support to, and remains ambiguous, about others. Many in civil society see the state strategically pushing the interests of the strong and neglecting the demands of the weak. Marks and McAdam (1999) point out that the political opportunity structure has been used to explain two principal dependent variables: the timing of collective action as well as the outcomes of movements. In addition, they underline that changes in legal or institutional structures that give more space to challenging groups are apt to set in motion the narrow and institutionalized reform groups that primarily exploit the new cracks in the system. This also means that organizers tailor their responses to changes in the political system that they seek to challenge, observing aspects that are most vulnerable or receptive to their efforts. With the FRA these cracks came in two forms: the election victory of the United Progressive Alliance (UPA), a coalition government that saw a rise in parliamentary seats for the Left Front, and the growing threat of Naxalism.

The UPA government came to power in an election victory, which many interpreted as a vote against "reforms" and, to a lesser extent, *Hindutva*.[16] Some social commentators (Gopalakrishnan 2006; Yadav 2004) saw this victory as a perceived setback to neoliberal reforms propagated by the previous government, the National Democratic Alliance (NDA). However, they also pointed out that there was no real change in power relations as "though widespread and deep-rooted discontent about the neoliberal policy package [reforms] exists, there is as yet no political formation that can focus such discontent beyond the regional and the issue-specific" (Gopalakrishnan 2006, 1). This was best summarized by Yogendra Yadav (2004), who pointed out: "The case that this was a mandate against policies of economic reforms is an overstatement. . . . [H]aving said this, it is equally necessary to realise that . . . if this election could [have been] a referendum on economic reforms, the policies of liberalisation would have been rejected." The only real shift was the rise in parliamentary seats won by the Left Front, which symbolized not so much a defeat for the neoliberals as it did the "almost but not quite" nature of their victory.

The immediate effect of this divide between the government and the Left Front was to create a kind of "institutional schizophrenia" (Gopalakrishnan 2006, 2) within the UPA government. While the structural policies of the newly formed UPA retained their neoliberal underpinnings, the government simultaneously initiated the process of creating the Common Minimum Programme (CMP), which became the basis for two new institutions: the National Advisory Council and the Coordination Committee with the Left parties. There is a possibility that this was done in order to achieve a balance that would retain neoliberalism as a political project, but with a human face. "The result was a government whose formal institutions were committed to neo liberalism but whose most vocal intellectual ones had a non-neoliberal, if not an anti-neoliberal, mandate" (Gopalakrishnan 2006, 3).

The second opportunity came in the growing specter of Naxalism, whose vision is one of replacing the idea of class domination or capitalism with a communist society, and its means to that end is by taking up arms against the state. The Naxals believe that "Congress administration represents the interests of the Indian feudal princes, big landlords and bureaucratic comprador capitalists" (Dasgupta 1974, 117). The justification of violence is rooted in the inevitability of the situation, the fact that the poor – Dalits and Adivasis, who are marginalized and subjugated – are not given access to a democratic forum to raise their concerns. Thus, the need to voice their grievances forced them to speak the language of violence. Second is the justification of "victimhood," where "violence was forced" on them, and hence, to save their land and their dignity, they were forced to use violence against the government (Kumar 2003, 4977–78). This use of violence was met with equally violent counter-reprisal by the state, resulting in more extreme marginalization of Naxal-affected areas. The state considers Naxalism a law-and-order problem rather than a movement, and its responses as a result have been security-centric. However, civil society and voices within the state did emphasize that a rights-oriented negotiated solution was (is) the need of the hour. When the UPA came into power, it saw forest rights as a tool to secure tenure rights, to counter the growing Adivasi discontent. It could use it as a rights-based solution, in addition to security centric activities, to keep the Naxal movement from spreading in the country (Ministry of Home Affairs 2007). Subsequent statements have underlined this:

> [A]dvanced artillery for police for the protection of local villagers along with a rigorous training of the Constitutional freedoms and protections of the citizens can satisfy the security concerns. Effective implementation of protective legislations such as the Panchayat Extension of the Schedule Areas Act, 1996 (PESA), National Rural Employment Guarantee Act, 2005 (NREGA) and The Scheduled Tribes and Other Traditional Forest Dwellers (Recognition of Forest Rights) Act, 2006 should be ensured.
>
> (Planning Commission of India 2008, 62)

Entering the deliberative space – finding political representation

By 2003, the mobilization around the FRA by the Campaign for Survival and Dignity and others gathered momentum. The National Democratic Alliance (NDA), just before the April 2004 elections, issued an order that the issue of tribal land rights should be settled within a year. This was strategically important as legislative assembly elections were to be held in some of the Adivasi-majority states – Madhya Pradesh, Orissa, Chhattisgarh and Jharkhand.

In 2004, the UPA government came into power with the support of the Left parties. Their Common Minimum Programme stipulated that the "eviction of tribal communities and other forest-dwelling communities from forest areas will be discontinued," and it stated that they would "reconcile the objectives of economic growth and environmental conservation, particularly as far as tribal communities dependent on forests are concerned and confer ownership rights in respect of minor forest produce" (Common Minimum Programme 2004).[17] Though there was initially resistance to formalizing aspects of these stipulations into law, there were some new spaces provided by the UPA government. The National Advisory Council was set up in 2004 as an interface with civil society and consisted of leading intellectuals and activists. This council was set up with the specific task of implementing the Common Minimum Programme of the government. It became the lobbying space for the coordinated movements of the NFFPFW, the CSD and other groups. This new institution, charged with its specific ideological focus "on social policy and the rights of the disadvantaged groups,"[18] was able to bypass the vested influence of a powerful Forest Department, overcome the neoliberal agenda of the government and highlight the need for a law to recognize the needs of tribal communities and forest dwellers at the national level in a series of high-level meetings with CSD and the government in the later part of 2004.

Soon after, in 2005, in a meeting chaired by the prime minister, there was official assent to two primary issues: (a) that a statutory recognition of forest rights was required and this would require a concise and clear law, and (b) the idea that since the forest bureaucracy had vested interests, it should not be in charge of any rights-recognition statute. Instead the responsibility to frame the law was transferred through an amendment to the Government of India (Allocation of Business) Rules to the Ministry of Tribal Affairs (MoTA). The MoTA was to become the nodal agency, which would consult with experts and activists. In the wake of the Supreme Court's involvement in encroachment cases and the MoEF handling of these cases, along with the paradigm shift from encroachment to tribal rights (Bose 2010), it was felt that the nodal agency in matters of framing laws regarding rights of tribal communities should be the MoTA, rather than the MoEF. The MoTA, however, did not have the institutional machinery, in terms of manpower or the reach of the Forest Department, which had serious consequences for implementation.[19] In addition the forest/mineral/tribal and poverty maps all overlap, which complicated issues by creating a more intricate web of vested interests (Bhushan 2008).

The MoTA constituted a Technical Support Group (TSG) to draft the bill which consisted of representatives of the MoEF, Law and Legislative Affairs, Social Justice and Empowerment, Panchayati Raj, Rural Development and Tribal Affairs, as well as representatives of civil society consisting of two environmental activists, two tribal rights activists and two legal specialists. Three members on the TSG represented the CSD. The CSD in the meantime drew up a draft law, many elements of which were included in the TSG draft. One of the major demands of the movement was the incorporation of non-tribal forest dwellers as a section, as well as state recognized Scheduled Tribes, in the ambit of the law. Tens of thousands of postcards were sent to the prime minister from villages all over the country, and more than 10,000 forest dwellers gathered in Delhi to demand the law's enactment (Ghosh 2005; CSD 2005). This attracted the attention of numerous parliamentarians and more than fifty members of parliament. It was estimated that around a quarter of a million forest dwellers and Adivasis rallied in demand for the act (CSD 2005).

Shankar Gopalkrishnan (2010) noted that although the expert committee included some strong provisions in the initial act, its dilution began once it passed the purview of the committee into the bureaucratic arms of the state. The forest bureaucracy termed the bill "too radical" and expressed concerns over its potential diverse impact on forest cover. In a letter to the MoTA, the MoEF questioned the very necessity of the Scheduled Tribes (Recognition of Forest Rights) Bill 2005: "In a strongly worded reaction to the draft, the ministry says the bill will destroy India's forest land and the failure on the development front should not be compensated by gifting away India's forest heritage" (Ganapathy 2005, 3). In response, Tribal Affairs Minister A. Kyndiah remained noncommittal, stating: "There is a strong view. . . . Let us see how it is resolved" (Ganapathy 2005, 3). In response, the prime minister's office called for a meeting in which the minutest details of the bill were examined, in consultation with all the concerned parties, and the conclusion was that the bill should not be pushed in haste (Bindra 2005).

Conservationists opposed it on the grounds that certain species of animals (such as the tiger) cannot co-exist with humans and demanded that some part of the forests remain inaccessible in order to protect these species. They also agreed with the forest bureaucracy that human habitation would cause depletions in forest cover. The bill also sharply divided the parliamentarian group. For instance, Tiger and Wilderness Watch was formed across party lines to work on one platform to conserve the tiger. The group included high-profile members of parliament (MPs) like Rahul Gandhi, Jyotiraditya Scindia, Jay Panda and Renuka Chowdhary, among others. Bindra (2005, para. 4) reported that "while most of the MPs were shocked at the Bill's impact, few came out in the open considering its political ramifications. An MP, who is presently a Union Minister, is reportedly worried that forests in one particular state, with only one tiger reserve, would be destroyed if forestland was given away to tribals in such a manner. . . . However, few came out openly to oppose the Bill, since it was considered politically incorrect." Bindra (2005, para. 6) also reported, "Tribal Bill is considered to be a master move of the UPA to win tribal votes. After the presentation, the MPs had a private discussion in which it was unanimously decided to steer clear of the Bill. Not one MP, whether from the Congress, BJP or the Left would publicly oppose the Bill as it meant loss of tribal votes, and political suicide."

A copy of the bill was eventually posted on the MoTA website in June 2005. The version was substantially different from the one that had been prepared by the TSG. The pro-bill lobby termed the bill "not radical enough" especially since it excluded forest dwellers, who were not counted in the category of Scheduled Tribes. In addition it placed an imposition of a cutoff date as 1980 and provided unclear provisions for relocation in reserved areas. Although it was estimated that around 2–3 million people were living inside India's protected areas (national parks and sanctuaries), there was no census of the number of Forest Dwelling Schedule Tribes residing within the core areas of national parks and sanctuaries (Madhusudan 2005). Therefore, it was not possible to calculate how much forest-land would be required in order to implement the provisions of the bill. The bill was eventually tabled in parliament in 2005.

Given the uncertainty about the bill and dilutions in the draft bill put out by MoTA, the movement took up a nationwide campaign to ensure that the draft bill was moved to the parliament as soon as possible. This campaign involved petitions, marches all over the country and planned protests to coincide with the winter session of parliament (CSD 2005). There was a protest march on August 15, Indian Independence Day, with the slogan "Nyaya Chahiye ya Jail Chahiye" (We want freedom or we want prison). Protests jointly organized by the Campaign for Survival and Dignity and the National Federation of Forest Peoples and Forest Workers were planned nationwide, courting arrest. National protests were also called by the Communist Party of India for "the tabling of the Bill and its passage with amendments" (CSD 2005).[20] The CSD, in their letter to the minister for Home Affairs wrote: "We write this letter to you to give you notice that, from November 15th onwards, we will begin a satyagraha[21] across India. We will protest, march and court arrest. We will fill the jails, for we would rather be prisoners in jail than hostages in our homes" (CSD 2005). In addition, the CSD also contacted members of parliament to garner support and tried to build bridges with moderate conservation organizations that were willing to discuss tribal traditions with regard to conservation.

In 2005, after being stalled in parliament, the bill was submitted for review to the Joint Parliamentary Committee (JPC), which was to systematically hear and consult with all the stakeholders and recommend appropriate changes. The Lok Sabha (Lower House of Parliament) speaker appointed a thirty-member JPC headed by Sh. K.C. Deo, which invited comments both orally and in the form of written submissions, but restricted only to Delhi because of lack of time. It heard representatives from all concerned areas between April and May 2006. Gopalkrishnan (2010) notes that as a democratically elected committee, the JPC were inherently less pro-bureaucracy and pro-state, which made it the ideal forum where the movement could advocate its position that had crystallized over the previous ten years. The movement organizations met all the members of the JPC and had the space for long discussions and debates. The NFFPFW, the CSD along with political parties of the Left, and activists pushed forward strong recommendations, which were unanimously accepted by the JPC in its report that was submitted in May 2006. The JPC shifted the 1980 cut-off year to December 2005, included in its terminology all non-tribal traditional forest dwellers, and recognized their rights in areas declared as protected areas. It revised the process for identification of such protected areas to ensure a more transparent process, and increased the ceiling of 2.5 hectares on land to 4 hectares but only after a three-level scrutiny into the veracity of claims. Most importantly, it prescribed that no diversion of forestland would happen without the consent of the *gram sabha* (the village assembly).

The JPC report elicited a resurgence of criticism against the bill from conservationists, forest department officials and officials within the MoTA. For instance severe criticism came (Ananthakrishnan 2006, 2) from the country's first-ever National Forest Commission headed by former chief justice of India, Justice B. N. Kirpal. The commission's report, submitted to the government in March, called upon the government to come up with another law, "providing the forest-dwelling communities a right to a share from the forest produce on an ecologically sustainable basis" (Ananthakrishnan 2006, 2). Blasting the bill provisions, it said, "[T]he politically motivated and ecologically suicidal proposal of providing temporary rights in these protected areas for a period of five years and if they are not relocated in that period that rights to become permanent, is a mere facade, and considering the past record and political motivations will never be achieved and the grant of such rights will irrevocably impair the ecological viability of protected areas" (Ananthakrishnan 2006, 2).

Others pointed out that the JPC's recommendations, when read together, could create an "explosive situation." Kothari (2006, 18) pointed out that there were potential pitfalls to the JPC stipulations, though they adhere to a human rights perspective. According to his analysis, the bill "no longer requires that the rights to the use of forest resources be sustainable. These omissions leave open the possibility of grave ecological damage in situations where the *gram sabha* may not be capable of stopping powerful inside or outside forces, or may not wish to regulate its own members from destructive forest use" (18). Other undefined terminology was the "expanded definition of 'traditional forest dwellers,'"[22] which "provide scope for State governments, land mafia and local elites to exploit the situation" (18). It also left definitions for basic amenities like roads open, and that could have serious repercussions in terms of ecological impacts in forested areas.

After the JPC made its recommendation, the bill became a site for a power struggle within the bureaucracy. A Group of Ministers (GoM) re-examining the bill issued a press release stating, "[T]he draft prepared by the Joint Parliamentary Committee would not serve the purpose of tribals and might even make them subservient to non-tribals in the forest areas." In addition, they felt that there exists a "spatial relationship between Scheduled Tribes and biological resources. This kind of relationship with nature does not exist in the case of non-ST forest dwellers" (Mukul 2006, 2). The GoM also pressurized the JPC to amend and water-down its recommendations to appease the conservationist lobbies. The JPC stood its ground, and protest demonstrations were held all over the country and in front of the parliament, calling for enactment of the bill. In a joint statement issued to the government, the organizations fighting for tribal rights stated: "We appeal to Parliament to accept the Bill in its JPC-amended form, as it implements the international environmental commitments and principles of forest biodiversity protection and protected areas in respect of the indigenous forest communities" (*Hindu* 2006, 1). Finally, the tribal and forest-dwellers organizations condemned the "anti-people and undemocratic stand of the Government on the Bill" (*Hindu* 2006, 1). According to Suhas Chakma (Asian Centre for Human Rights), "[T]he tragedy of the bill was that we were not fighting for anything new, we needed a law to circumvent the Supreme Court order" (personal communication, April 22, 2009).

The CSD and other organizations in the face of this standoff began pressuring the government to enact the law. The recommendation of the JPC had given legitimacy to their claims. Communist Party of India (Marxist) called for mass protests on July 18, 2006. The CPI(M) Maharashtra state committee decided to propagate the importance of the changes made in the bill and the amendments to the proposed bill accepted by the parliamentary committee. Three thousand booklets and 25,000 handbills were distributed. In some places, *jathas* (rallies) were started (Shiralkar 2006). The beginning of the monsoon session of parliament (November 22, 2006) saw the CSD and other organizations call for a massive *dharna* (sit-in), demanding "immediate passage of this legislation with the required amendments. We also demand the repeal of these and other policies that will result in massive environmental damage and the loss of our homelands" (CSD 2006). In addition massive rallies were held in the capital cities of Bangalore, Bhubaneswar, Mumbai and Ranchi (CSD 2006).

In December 2006, the law was finally passed in parliament but only after some hurried modifications, leaving the ministers of parliament with insufficient time to discuss or analyze them. Some pro-JPC quarters were able to infuse some of its spirit into the bill, retaining the inclusion of non-tribal communities, the new cut-off date and the new procedure for resettlement. However, it disregarded the key role played by *gram sabhas* that kept control over decision-making. This left the forest bureaucracy in control, open to the influence of conservationists and, some

Table 4.2 Process of notifying the FRA (source: Upadhyay 2011)

TSG on act-first draft
Draft published with minor corrections
Joint Parliamentary Committee (JPC) – JPC Draft
GoM-negotiated draft
The Forest Rights Act (published for information on January 2, 2006)
TSG on rules-headed by Sh. S. R. Sankaran (former bureaucrat)
Draft revised by MoTA
Further "refined" by law ministry
Draft pre-published as required by the act on June 19 for forty-five days
Comments received through August 7–8, 2007
Final notification on December 31, 2007, and rules on January 1, 2008

felt, mining and business lobbies as well. One member of parliament commented that he "welcomed the Bill but criticised the Government for not bringing it in the right manner since the members had received the amendments at the last minute. He said the Government did not seem comfortable with the tabling of the Bill but was doing it under pressure. Instead, the Government wanted to protect the tigers" (*Hindu* 2006, 1). The bill that became the Scheduled Tribes and Other Traditional Forest Dwellers (Recognition of Forest Rights) Act of 2006 finally passed in the parliament on December 15, 2006, the last day of the winter session.

The role of civil society: the conflict of narratives

In relation to the contestation over forest rights, Gopalkrishnan (2010) identifies three different actor constellations entrenched in three different social goals. These are (a) the forest bureaucracy with its allies in hardliner conservationists and the English press in India; (b) non-forest state bureaucracies, NGOs and some progressive elements within the forest bureaucracy; and (c) forest dwellers, tribal movements and some parts of the Left-leaning political parties. The first group demanded that the status quo be left in place, calling for "centralization," "autocracy" and the "enclosure system of forest bureaucracy." The second group saw the struggle as "legitimate but limited." It focused on the immediate problems and wanted the state to accord the necessary rights to indigenous people with focused legislative effort, bringing the situation under control without disrupting the deeper structures of resource control. It wanted to bring stability to a country embroiled in severe internal conflict. The third group demanded nothing less than a complete overhaul of the existing status quo, bringing empowerment and voice to the most marginalized in the Indian state.

This contestation played out in the different levels of the struggle for the act depending on which actor had most influence over each stage of the political process. The key areas of contestation would be the main themes of the law, which were the following:

1. the list of rights, including right of occupation, non-timber forest produce (NTFP), grazing rights etc.;
2. determining the definition of who is a "forest dweller" and therefore entitled to these rights; and
3. creating institutional frameworks to recognize these rights and delineate inviolate forest areas and conservation/wildlife protection.

In the weeks leading to the passing of the FRA, debates on the advantages and disadvantages were heated with several entrenched interests articulating debates surrounding the bill.

Conservationists and forest bureaucracy – the narrative of ecological modernization

Rights in the FRA are linked to a wider managerial discourse. One of the last-minute amendments to the FRA appeared to be aimed at ensuring that the allocation of rights did not lead to unsustainable use of forest resources. As we saw in the previous sections and chapter, the discourse of the neoliberal state is increasingly one of eco-modernization with emphasis being placed on the management and valuation of ecosystems. While the FRA no longer mentions that forest dwellers are responsible for managing the forest, mention of this is found in the preface, and the centrality of "sustainability" is reiterated (Menon 2008, 10). The responsibility in keeping forests sustainable is also emphasized in the preamble vesting forest dwellers with "the responsibilities and authority for sustainable use, conservation of biodiversity and maintenance of ecological balance . . . thereby strengthening the conservation regime of the forests while ensuring livelihood and food security" (FRA 2006, preamble).

The oppressive control and constraints that the forest bureaucracy placed on indigenous people had been reiterated throughout the struggle for the FRA. Their strict enforcement of "fortress conservation" was given credence by highly vocal hardline urban wildlife conservationists popularly dubbed the *tiger-wallas*. This group of forest bureaucrats and hardline conservationists were severely opposed to the act. In the 1990s, many environmentalists believed that wildlife conservation required inviolate "wilderness" areas and suggested that the co-existence between humans and wildlife was not possible. The Tiger Task Force report of 2005 referred to previous policies of "people versus parks" – and the inevitable corollary, "people versus tigers" (MoEF 2005, 140) – and critiquing the state, it came to this conclusion: "One, the conservation regime rededicates itself to a command-and-control mode of wildlife preservation. Two, it becomes no longer necessary to refer to or think of 'people' while speaking of or planning for conservation" (MoEF 2005, 141). The authors of this report were certain that the approach of "guns, guards and fences" is simply not the answer; however, the report did insist that there needed to be "expansion of inviolate spaces for the tiger by minimizing human pressure in these areas" and that regeneration of the forest habitats should be at the fringes of the tiger's protective enclaves by "investing in forest, water and grassland economies of the people" (MoEF 2005, 146). Much of the debate around the act was centered on environmental consequences. The concerns highlighted before the bill turned into the act included the lack of territorial limit on claims to forest resources (Mohanty 2005, 33) and the fact that rights to land in protected areas would be detrimental to wildlife because wildlife does need large tracts of uninhabited land (Madhusudhan 2005, 4894). These were addressed by the state, resulting in what many considered a dilution of the bill. It has been pointed out that this dilution was despite the creative lobbying done by the organizations representing the tribal communities, and possible reasons for the dilution is that wildlife conservation is central to the elite and middle-class imagination and influences national environmental politics.

The protected area conservation approach has led to years of forced displacements from "core zones" and also to limiting or suspending communities' access to natural resources within these zones. However, it has also kept last tracts of ecologically sensitive areas protected as well as conserved many threatened species of wildlife. Conservationists sympathize with the plight of the people but insist that in order for quality of forests to be maintained, zones need to be demarcated through a "command and control" approach and the responsibility should lie solely with the Forest Department. The campaign run by conservationist lobbies against the FRA culminated, for the first time in India's history, into campaign ads running against legislation. It also led to forest bureaucracy and conservationists filing nine court cases in an attempt to get the Forest Rights Act struck down.

The debate over forests rights was framed as "tiger versus tribal." Through the influence of the Forest Department, the GoM inserted the clause of "Critical Wildlife Habitat" (CWH) into the act. Recognition of rights in wildlife sanctuaries and national parks was one of the most contested provisions of the act, with conservationists as well as the MoEF demanding that all protected areas be kept out of the act's ambit. Rights activists, on the other hand, pointed out that many of the protected areas had been notified arbitrarily without any concern for, or consultation with, the people living within them and that there was no scientific evidence that co-existence of people and wildlife was not possible in most cases. All the same, the MoEF issued guidelines for the early identification of CWHs (and Critical Tiger Habitats [CTHs]) even before the act had come into force. It finally had to abandon the notification of CWHs due to a member of parliament charging the secretary of the MoEF with breach of parliamentary privilege as CWH has been defined only in the FRA and such areas could not be notified before the law had come into force (Sarin and Springate-Baginski 2010, 27).

The CWH included the concept of core areas, which existed as a management concept; areas within sanctuaries and "inviolate areas" were often converted into national parks in order to keep them out of bounds for communities. However, the clause within the FRA mentions concepts such as "irreversible damage," "coexistence" and "inviolate" without defining them. Some conservationists argue that these concepts are necessary to keep part of the forest pristine and that scientific research has supported that areas within the forest with human habitation were far more denuded than the areas kept out of bounds. Now the same core areas, with the new terminology of "Critical Wildlife Habitat," were given legal backing. The determination of these habitats would be done as per MoEF guidelines, and decisions pertaining to relocation or issues of coexistence were the responsibility of the Forest Department. "More importantly, these huge areas – whose boundaries would take months to rationally decide even within a single tiger reserve – are to be mapped out in *all* tiger reserves within 30 days" (CSD 2006). The tiger task force has pointed out that the government should "take into account the options for livelihood in the resettled village. . . . The relocation package must be designed to provide viable alternatives" (CSD 2006). The act also emphasizes that state relocation must provide "secure livelihood." Major environmental organizations such as WWF-India, Kalpavriksh, Foundation for Ecological Security, Samrakshan Trust and the Vidarbha Nature Conservation Society, as well as some of India's top conservation scientists, had previously opposed this kind of ad hoc process in the name of relocation. In March 2006, in a joint statement, these groups and the CSD had said no resettlement should occur without "a site-specific open process, with involvement of . . . multi-disciplinary experts, [which] should take place through democratic mechanisms including local community representatives."

Otherwise there would be "legal battles, physical resistance, and enhanced conflict. This will seriously harm both conservation and people's rights" (CSD 2006).

This reliance on experts and exclusionary policies is evocative of the historical preferences for eco-development and habitat management of the forest bureaucracy. The economics of this control is particularly revealing – for example the FRA challenges the forestry administration's monopoly over forest management by giving forest-dependent community groups the statutory right to collect and sell non-timber forest produce (NTFP). This threatens the state forest department's control over NTFP trade worth over US$10 billion a year (Mahapatra et al. 2010; Sethi 2011).

Social movements and the articulation of rights – the narrative of equity

The historical frames of injustice and exclusion from India's constructed forests fed into the narratives of rights activists. The narratives of dispossession emerged through the works of anthropologists like Verrier Elwin (1939) and Fernandes and Menon (1988). Modern ethnographies of grassroots movements contextualized narratives in data and analysis and put them in a larger socio-economic context of post-independent India (Sarin 2003; Garg 2005). These seemed to feed into the crystallization of a movement that came to a head with the 2002 eviction orders. The *Jan Sunwai* (public hearings) of 2003 that brought forest dwellers together in a shared experience of dispossession was collected in a book (*Endangered Symbiosis* [CSD 2003]). Thus drawing from the history of exclusion, it set out to unify the image of the tribal as "one who lives in harmony with nature," of "unbroken heritage" with a "special relationship to the land," and who required "constitutional protection."

This bloc who constituted the tribal movements argued that marginalized groups could use the democratic political apparatus to "defend the material basis for their moral economy against enclosure and capitalist (and state) accumulation, and even reverse it" (Sarin and Springate-Baginski 2010, 3). They questioned the contention around this act affecting the poorest and most dispossessed when corporations and powerful actors with vested interests subverted environmental laws with far greater ease and remained unchallenged. This rights-based coalition discussed and proposed several safeguards to keep the act from elite capture.

Initially this narrative simply highlighted the needs of the forest dwellers, living in the fringes of society, attempting to organize them and give them voice. After the critical juncture of 2002, this concretized to more definite claims over land, community tenure and rights over resources, rights to habitation and cultural rights like the right to practice shifting cultivation (Suhas Chakma, personal communication, April 22, 2009). Several frames were reiterated and disseminated effectively through media, presentations and protests. Some of these were the Forest Department as an oppressive *zamindar* (landowner) and conservationists as "elite" – "who clearly had vested interests in the name of tigers" (Ashok Chowdhury, personal communication, May 17, 2010). In addition movement organizations attempted to flesh out the idea of representation. They encouraged the Adivasis to "represent themselves because we are the worst [people] to represent them" (ibid.) in dialogues with the state or Forest Department. However, they found that "administration and us [movement leaders] speak the same language, fight in the same language and are generally comfortable. They don't speak to the adivasi" (ibid.).

In addition, they came in direct confrontation with alternate narratives. In their struggle against the conservationists, they wrote several open letters to various well-known wildlife groups, including the Bombay Natural History Society, Wildlife Protection Society of India, Wildlife First, and Conservation Action Trust. The CSD and NFFPFW invoked the International Union for Conservation of Nature "Durban Accord and principles of the Convention on Biological Diversity" to urge conservationists to start an "open dialogue" and put their faith in the "democratic approach to conservation in which both wildlife and people's rights are given importance" (NFFPFW Convention 2009, circular). They particularly attacked the NGO Vanashakti[23] for not being transparent and not disclosing who funded the TV campaign that it ran against the FRA. They charged Vanashakti for being "inaccessible to the oppressed and opposed to their interests." Using the discourse of revolution, they referred to "urban conservationists" who remained ignorant of grass roots realities (CSD n.d.). These conflicting values manifested in some key policy demands and lead to much debate in the process of ratification of the FRA. Some of these key ideas are set forth in Table 4.3.

There were also "a mixed group of people both from within and outside government, who supported the law but did not favor a major change; they wanted only limited recognition of land rights, just sufficient to ameliorate the underlying causes for the intense conflict prevalent in forested tribal inhabited areas" (Sarin and Springate-Baginski 2010, 6). This group – consisting of a mix of state bureaucrats, conservationists and NGOs – took a neutral position on the act, expressing concern about certain provisions within it. The commercial lobby remained curiously silent during the debate and has been accused by many of hiding behind the conservation lobbies. This mixed bloc argued that the cutoff date of December 2005 made it more challenging to determine the veracity of the claims. They also voiced concern about the possibility that this act could lead to further fragmentation of the forests as communities cannot be cut off from development, and roads, school and hospitals would have to be built. They pointed out that, thus far, living within the forests without access to the basic development infrastructure has contributed to an unbroken traditional lifestyle but that there is a possibility that, with the act coming into power, there will be a breakdown of traditional lifestyle in the long run. There were also concerns about misusing the act – for instance encroachers who could use the act to swallow up land for industries who could subvert the act to use land allotted to tribal communities for commercial purposes. For many within this group, the tribal bill was an extension of the 1970s slogan of the Socialist Party, "Jhaad hamarey, zameen hamari" (The trees and the land belong to us), leading to complete destruction of forest in the Jhabua and Dhar district (Singh 2005; Buch 2005).

Table 4.3 Points of conflict (adapted from Upadhyay 2011)

Tiger advocates views	Tribal advocates views
• Inviolate areas should be established for flagship species like tigers (4.3 percent PA). • With the recognition of rights, forest land will eventually pass on to rich non-tribals defeating the purpose of the act. • With the transfer of hundreds of thousands of hectares to tribals, the last remaining forests will be destroyed. • Tribals, other traditional forest dwellers and *gram sabhas* do not possess the technical skills to manage and protect forests. The Forest Department has these competencies and thus must retain some control. • The act is primarily based on electoral politics and the regularization of encroachment.	• It is "historical injustice" to those communities most dependent on forests for their livelihood to not grant them access to land that is rightfully theirs. • Tribals and other traditional forest dwellers have been left out of the decision-making process in managing the forests despite the overwhelming evidence of deep knowledge of local management practices. • Exclusionary conservation practices have impoverished local communities. • Security of tenure is a key element to involve tribals in forest management. • Rights of these marginalized people have been revoked without any legal due process. • Administrative laxity continues due to state apathy towards such communities.

Others pointed out that conservationist research contains a bias towards attributing human harm to ecosystems, even when there is no evidence to support this hypothesis. Rather, evidence has pointed to the contrary. For example when cattle were banned in the Bharatpur national park in 1982, the park's habitat went on slow decline (Lewis 1995). Examples like these give credence to their views that forests in India have evolved in the context of interaction between humans and biodiversity. Activists have pointed out that forest governance has been shaped in a certain way for hundreds of years without evident benefits, and it is time for a new human-centered approach to conservation. The Forest Rights Act, they argue, is well placed to do this given that there are inherent checks and balances within the act to ensure sustainable practices. For example, three levels (village, sub-divisional and district) of consent are required before any diversion of land. They believed that the FRA could play an important role in making communities true stakeholders of conservation.

It was argued that the act empowers the Adivasi communities to protect their forests. This was the case in 2008, in the Karlapat Wildlife Sanctuary, where local communities from Orissa, empowered and encouraged by rights accorded to them by the act, seized three truckloads of timber from the residence forest range officer. Earlier, villagers said, they had failed to curb timber smuggling as they did not have any right or say in the management of resources inside the sanctuary. It provides another avenue to fight against some of the hugely destructive "development" projects. For instance, the act has already been used to stop mining of bauxite in the thickly forested and sacred Niyamgiri hills of the Dongaria Kond tribe in Orissa. In contrast, the Karnataka State Forest Department has drawn up plans to move the tribe of Soligas out of the Biligiri sanctuary, though they have coexisted with the tigers for centuries, but have not revoked the leases granted to large companies for 1,800 acres of commercial coffee plantations inside the sanctuary. Thus, many believe that tribal communities and participatory management of forests will work more effectively in the long run because of the sense of ownership that the FRA accords local communities.

Conclusion

The Forest Rights Act aimed "to recognise and vest the forest rights and occupation in forest land in forest dwelling Scheduled Tribes and other traditional forest dwellers who have been residing in such forests for generations but whose rights could not be recorded; to provide for a framework for recording the forest rights so vested and the nature of evidence required for such recognition and vesting in respect of forest land" (FRA 2006, preamble).

The FRA as it stands is a landmark victory for the struggle that aimed at a complete overhaul of the conservation and land rights paradigm. One has also seen a marked weakening of the act through debates and dilution. The dilution of the act is clear in the provisions of local-level *gram sabhas* and the fact that much of the revenue collected from the non-timber forest produce is still channeled through the state, which is still not bound to provide any minimum price. The implementation of the act is still encountering opposition from the forest bureaucracy and conservationists in diverse regions in the country. The act, however, does represent a democratic momentum that is diffused through the bureaucratic structures of the country. The diverse voices, entrenched in different value systems and visions, were able to influence different aspects of the law at different junctures of its formation, and in itself the process of ratifying the FRA represents the struggle for consensus through both consultative means and social movements. In highlighting the hegemonic conflicts between the different visions of development and the struggle for rights, it was a landmark piece of legislation that attempted to correct a "historic injustice," and, through its participatory process, it exposed the structures on which development policy is based. Only the process of implementation will tell whether it was merely a symbolic victory or a true shift in the politics of the state.

McAdam et al. (1988, 709) suggest two functions in the mobilization process that are both fulfilled by this case study. The Forest Rights Act provides the context for realization of collective action by civil society organizations, with most of their demands being framed in the context of laws and rights. The key success of this process remained in mass mobilizations around the frames of tribal communities as the "dispossessed" and "protector of the forests." The idea was that the movement would restore dignity to those who had suffered historical injustice. In addition, the different processes that the FRA underwent provided a staging ground where the different points of view were deliberated on, sometimes heatedly and at odds with each other, but all represented. The NFFPFW and CSD advocated for the role of the marginalized, and state forest bureaucracy and hard-line conservationists advocated for the position of forests and tigers, which they considered marginalized in the discourse about people. They followed the process of both "softening up" and "coupling," interpreting specific interests in light of larger public concerns and taking advantage of political opportunities throughout the process.

Though it is commonly believed that the demand for forest rights came from the 2002 eviction drives, this chapter shows that integrating discourse, that for decades was only heard at the margins, into state policy required long years of forum organization that eventually was backed by the strength of the voting masses. If this had been a spontaneous response to eviction drives, it could have simply frittered away, and the government's liberal agenda would have persisted. In contrast, the building up of structures of mobilization over decades allowed organizations to tailor their responses to political opportunities provided by the state, entering forums of deliberation when required and using extra-parliamentary tactics of mobilization and public pressure when the government threatened to retreat. The success of deliberative politics, therefore, depends on non-institutionalized public spheres that acted as a "context of discovery" (Flynn 2004). The informal and diffused nature of public deliberative spaces, close to the grassroots level, allowed for the process of framing the problems that require redressing by the formal political system. These could then be legitimized by mobilization of people around these frames.

Civil society, with its multitude of faces, played a role in different parts of the policy process. Both social movements and conservationists created interfaces with the state to lobby for certain provisions, finding spheres of influence both within the National Advisory Council and at the deliberative space provided by the Joint Parliamentary Committee. The movement allied with the Left Front to find support within the structures of government and looked to building mass mobilization to give legitimacy to its claims to the point that to publicly voice dissent was considered "political suicide" by members of parliament. One highlight of the NFFPFW was that it encouraged tribal leaders to step to the forefront of the struggle to change the language of negotiation. One NFFPFW coordinator explained that the departments of the government were clearly uncomfortable in negotiating with people who were from a completely different background. They used a different vocabulary from the one the NGOs, whose members had been schooled in urban areas and had the same background as the bureaucrats they negotiated with, used. In contrast, wildlife NGOs built their strategies around the institutional and bureaucratic machinery of the state, and in particular, the Forest Department, which was their closest allies in this process and will be crucial in the implementing of the act. In this scenario, mobilization did not just promote collective action but also performed a complementary role of political communication as a deliberative legitimation process (Habermas 1996). I have interpreted this communicative power to the discursive power produced through arguments and narratives disseminated both historically and within informal public spheres. Communicative acts in this case must be understood as a struggle for power between opposing groups (Mouffe 2005) or as directly transforming existing democracy (Hardt and Negri 2000). This communicative element is particularly significant as it widens the narrower definitions of deliberative spaces, encompassing also the language of civil society.

The history of democratization demonstrates that pressures for greater democracy are almost always "oppositional" rather than purely deliberative (Cohen and Arato 1992; Fraser 1990; Tarrow 1998; Dryzek 2000). Coupled with oppositional discursive frames, collective actions such as sit-ins, strikes, radical demonstrations and urban riots "can create fear of political instability and so draw forth a governmental response" (Dryzek 2000, 101). Thus, the key achievement of the process of enactment of the Forest Rights Act is in the utilizations of deliberative spaces coupled with the use of mass mobilization to draw forth a government response when those spaces were threatened. This ultimately led to a more organic form of participation, creating a policy that enjoyed legitimacy and reflected the demands of the people who were historically more marginalized.

Notes

1 A discursive frame is the set of cultural viewpoints that informs the practices of a community of social movement organizations. Each discursive frame provides a cultural viewpoint from which the environmental organization acts. This discursive frame defines the goals and purposes of the organization, and provides guidance for the actions of the organization (Brulle 2008, 3)
2 This was later supported by statistical analysis indicating that more than 50 percent of the variance in the loss of forest land can be accounted for by the amount of profits state governments generated from the forests (Khator 1991,137).
3 Government land, but does not require acquisition of rights, nor the curtailment of activities as envisaged in National Parks and Wildlife Sanctuaries.
4 A private or community-owned area where the community has voluntarily engaged in wildlife conservation, and requires financial and technical assistance for its future management.
5 Study conducted by the Centre for Equity Studies in New Delhi, in 2004.
6 Control over management of the forests.
7 "Armed Conflicts Report – India-Andhra Pradesh," www.justice.gov/sites/default/files/eoir/legacy/2014/02/25/India_Maoist.pdf (accessed September 9, 2010).
8 *Down to Earth*, "Interview between T N Godavarman Thirumulpad and Surendranath C," August 31, 2002.
9 Friend of the court: one who volunteers to offer information to assist a court, in the form of either a legal opinion, a brief, a testimony or a learned treatise on a matter that bears on the case. The decision on whether to admit the information lies at the discretion of the court.
10 Order in I.A. no. 703 filed by amicus curiae highlighting serious problem of encroachments on forestlands, August 12, 2002.
11 Ministry of Environment and Forests press release, May 10, 2005, minister of Environment and Forests; statement regarding steps taken by government for regularizing forest land being cultivated by farmers in Lok Sabha.
12 The labor relations in Adivasi societies organized around labor, community and collectivism but not exploitation of labor nor in selling its own labor.
13 Organizations that played a critical role in this process were BIRSA and Judav (Jharkhand), NESPAN and Forest Villagers Union (northern Bengal), Vasundhara

and Tenduparta Mazdoor Union (Orissa), Yakshi (Andhra Pradesh), CORD (Karnataka), Tribal Development Society, Chengalpett (Tamil Nadu), Nagrik Manch (West Bengal), Van Kamgaar Union and Shashwat (Maharashtra), Ekta Parishad (Madhya Pradesh), SAVE (Himachal Pradesh), Jan Mukti Andolan (Kaimur, Bihar), Vikalp, Van Tongia Vikas Samiti (Gorakhpur, Uttar Pradesh), and Tree Grower Cooperative (Gujarat).

14 The order of the MoEF dated May 3, 2002, addressed to the Chief Secretaries, Secretaries (Forests) and the Principal Chief Conservators of Forests of all States/ Union Territories, available at www.pucl.org/Topics/Industries-envirn-resettlement/2003/forest-rights.htm.

15 National Forum of Forest People and Forest Workers Pamphlet, *OpenSpace*, http://openspace.org.in/node/416 (accessed June 19, 2010).

16 *Hindutva* or cultural nationalism presents the Bharatiya Janta Party's conception of Indian nationhood. It must be noted that *Hindutva* is a nationalist, and not a religious or theocratic, concept.

17 http://pmindia.nic.in/cmp.pdf (accessed April 13, 2012).

18 http://ssa.nic.in/rte/National%20Advisory%20Council%20under%20the%20RTE%20Act (accessed August 19, 2015).

19 "A serious challenge being faced by MoTA is the lack of adequate human resources to monitor implementation of the Act." A.K. Srivastava, Director of the Ministry of Tribal Affairs (MoTA), presented the status of implementation of community forest rights at the Future of Conservation Network's meeting in New Delhi on August 17, 2009.

20 Campaign for Survival and Dignity Forest Rights Update: Call for All India Protests, November 7, 2005 (personal communication).

21 Marching for truth.

22 Including communities resident in or adjacent to forests for at least three generations, people forced to occupy forest lands after being displaced from development projects and not being provided rehabilitation, or people forced into "encroachment" owing to other failures of the government.

23 Literally meaning "power to the forests," it was a conservation NGO that ran a campaign against the Forest Rights Act through websites and TV advertisements.

5 Conclusion

The past decade has seen a growing demand for better environmental policy formulation and enforcement by environmentalists and civil society. At the same time, the environmental reform process has been viewed as regressive or lacking teeth in creating accountability mechanism for both industries and regulators. To counter this, civil society has called upon the government or used courts at every opportunity to repeatedly ask for meaningful participation. The only policy that enshrines this formal space for participation is the Environment Impact Assessment Notification 2006 (amended from the earlier notification of 1994). Even this is a token attempt by the state, limiting the participation to those who have "plausible stake in environmental aspects of the project." In spite of this, there have been some successful processes for inclusion in India, like the public consultation process held in seven cities on the introduction of BT Brinjal (2010). In general, one finds that other than the public hearings, invited spaces in environmental policy formulation are almost non-existent. In fact, even the role of public hearings and consultation remains precarious, as evidenced by a memorandum in July 2014[1] that exempts public hearings for coalmines of less than 16 million tons per annum. In short, much of the participation in policy formulation and implementation has become mere symbol, because of a lack of engagement and information by the state, willful narrowing of access to deliberative structures by project proponents, illiteracy and marginalization of affected parties (Patra and Satapathy 2014).

In the opening chapters I discussed how traditional, hierarchical policy processes within the Indian state have given way to more deliberative democratic norms. These democratic norms of deliberation have been subject to much criticism. The lack of empirical evidence supporting deliberative democratic claims has been identified as a major lacuna. Many theorists have challenged the approach's insufficient treatment of power, politics and the value conflicts that are inherent in discussions on environment and society. Furthermore, questions have been posed about the eking out of such spaces of deliberation within the policy context, especially in developing countries. However, at a rhetorical and policy level, the idea of a deliberative democracy has gained traction: for example in 2013, the National Advisory Council (NAC), led by Congress President Sonia Gandhi, had recommended mandatory pre-legislative consultation for all proposed laws and rules. It wanted to "create institutionalised space for people's participation in the formulation of legislations in a systematic manner." It had also said that such a policy of consultation would take the country from a "representative democracy to a participatory, deliberative democracy.[2]

The case studies presented attempt to reconcile these two diverging trends of deliberation – political currency at a rhetorical level and dilution in the real world – by presenting interaction of two processes with the state. The first, centered on biodiversity, portraying "invited spaces" that had been critically acknowledged as uniquely deliberative, is considered a failure in terms of formulation with the state. The lack of will from elites, the power asymmetries between key social actors, the relative value placed on expert knowledge and the characteristics of the state have been discussed in order to underline that participatory norms are often meaningless. Instead they are used as a way of placating civil society, which demands more substantial and visible reform. In contrast, "invented spaces," resorting to modes of protest and mobilization in order to reinvent spaces of participation, can lead to more effective deliberation within the policy spaces or forums provided by the state. The more organic structures of this mode of participation ensure a high level of engagement of citizens throughout the process.

Each case study is a critical narrative that demonstrates the role of political opportunities and the actions of civil society in raising or lowering the influence of public participation on decision-making in potentially deliberative processes. This conclusion highlights the similarities and differences between the two case studies that reveal the difficulties and complexities that must be addressed if the goal is to achieve inclusive processes of deliberation for environmental decision-making. The objective of this research was to understand what strategies of advocacy are utilized with deliberation at the national level that impacts policy changes. These cases reveal how macro-level political realignments – both global and national – forced new institutional provisions for participation. In doing so they attempt to answer some broad questions:

1. How do civil society organizations (CSOs) respond to different policy openings?
2. Under what conditions have participatory spaces emerged?
3. What roles do narratives play?
4. What does this imply for the practice and theory of deliberative democracy?

Comparative elements

How do CSOs respond to different policy openings?

The role of civil society, outside of the state, as a sphere for informal deliberation is particularly important in forming cohesive demands that are then presented in formal deliberative, empowered structures. This might be referred to as a "public sphere." Micro theories of deliberative democracy carve a space for civil society within structured deliberative forums. They suggest that civil society actors should engage in communicative forms of action through collaborating with the state. In contrast, macro theories of deliberative democracy emphasize the informal and unstructured nature of public discussion, where civil society is called upon to play an informal role both outside and in opposition to the state, which require both communicative and strategic behavior (Hendriks 2002). In relation to the case studies, these neat cleavages are collapsed through the advocacy strategies utilized by civil society actors to manage their relationship with the state. Invited spaces of deliberation require actors to maintain close links with the state and bureaucracy in order to facilitate influence in different organizational structures within the state. However, in order to represent the interests of those marginalized by the state, strategic links have to be maintained with the informal sphere where demands are negotiated and collated by those who have less access to the formal institutional spheres of the state. In contrast, social movements which grow from macro deliberative spaces, in opposition to the state, also have to be present to negotiate demands within micro deliberative spaces – like state and civil society forums, steering groups and committees – to negotiate their demands. In contrast to Habermas's (1996) idea of the "two-track" model of democracy, where deliberation proceeds on two levels, in the cases observed here, opinions are formed in the public sphere and then transmitted via "currents of public communication" to the state where more formal deliberation takes place in courts and parliaments for the purposes of "will formation"(lawmaking) (Habermas 1996, 307–8). I suggest that maintaining the links between both the micro and macro deliberative sphere through strategies of negotiation and opposition at critical junctures by democratic deliberators form mechanisms of transmission and keeps the state and civil society more porous to those same currents of public communication.

Grassroots movements and the demand for substantive rights

The micro-movements in India represent many different arrangements. The literature on movements describes them as grassroots movements, social movements, non-party political formations or processes, community-based or mass-based organizations, social-action groups and movement-groups (Sheth 2007). This study uses these terms interchangeably, but the reference is specifically to a particular genre of social movements, which gained momentum in the mid-1970s in India. These movements' aim was to democratize development and transform the overriding power relations in society (Kothari 1989; Sethi 1984; Sheth 1984). The decline of mainstream institutions of representative democracy – the legislatures, elections, political parties, and trade unions (Sheth 2007) – as a space for the articulation of collective rights allowed movements to consolidate outside the representative political space. Given this idea of movements opposing bureaucratic structures, CSOs that enter partnerships with the government are often seen as coopted, losing their independent stand. However, the NBSAP process showed that CSOs do not always have to take a stand in opposition to the state in order to highlight issues of equity and justice. Often constructive arrangements with the state can nurture emerging spaces of articulation for those sections of society without voice.

Both the NBSAP and the FRA had different strategies and agendas. The NBSAP was an indirect movement, a political process that appropriated networks and movements, including self-help groups, indigenous people's movements, farmer seed-sharing associations and so on. Though the TPCG stood for indirect engagement, representing people who were directly affected, they saw their role more as "building bridges" (Apte 2005, 135). "[Kalpavriksh] acted as an information dissemination agent, supported the struggles of those on the ground, helped organize yatras and exchange programmes involving diverse stakeholders, and organized a series of national consultations on the subject" (Kothari 2002, para. 8). However, they were activists, building a movement from the top-down. The focus of the movement was to create awareness and empowerment, allowing the voices of the grassroots to be legitimized in national policy. This aim was not too different from the focus of the mobilization around the Forest Rights Act. However, the mobilization around the FRA aimed to involve direct stakeholders, those who were negotiating the adverse impacts of economic changes in their own communities and livelihoods. The problems of direct representation in a deliberative set-up with the state have been mentioned in the case study. There we saw that state officials were uncomfortable with the unfamiliar vocabulary and articulation of rights. One interviewee mentioned (TSG member, personal communication, October 3, 2011) how many of the concessions to the movement could only be decided in the deliberative space and with the support of intermediaries, trusted by both the government and the social movement organizations.

Another characteristic in both processes was the link between urban and rural, both in terms of structural causes of poverty and in terms of networking. Actors within these processes highlighted that root causes of loss of voice, access and livelihoods are linked to patterns of urban growth that India has followed. They highlight global processes and their impact on the marginalized, naming structural adjustment programs, international donor mandates and globalization, among other processes, that have led to further marginalization of resource-dependent communities. This has, among other thing, led to networking and building of new organizational linkages between the city-based and village-based social action groups. This has allowed for a three-pronged approach where grassroots activities are undertaken to disseminate and collect information and knowledge, deliberate at different levels of government and give people far from urban centers of governance a voice that is reflected in planning processes. These networks derive legitimacy from the grassroots in order to represent their issues in deliberation at the national level. The key failure of the NBSAP process was the inability to garner the grassroots support at the point where deliberations reached a stalemate with the state. This cast aspersions on the overall legitimacy of the process.

"The concept of 'grassroots' was once very specific: it meant the basic building blocks of society – small rural communities or urban neighborhoods where the 'common man (or woman) lived'" (Batliwala 2002, 396). This term was used to signify the poor, laborers or the working class, as opposed to dominant social elites, and it usually applied to rural, village-level communities rather than to urban ones. Today the context of the grassroots has changed and so have the social movements clustered around it. The differentiations between global civil societies and ones operating within the borders of a particular state are significant. The terms "grassroots" and "grassroots movements" disguise the very real differences in their power, resources, visibility, access, ideology and so on, and often act as symbols and microcosms of the larger power imbalances between northern and southern power holders. The focus of this study has been on different processes operating within the nation state. The influence of global rights-based discourses have been subtle and played out in the processes around the NBSAP and the FRA. Conventional legal notions of civil liberties have given way to the expansive ideas of rights, where groups are evaluated on the basis of their access to basic needs, articulated in terms of rights and entitlements. Grassroots movements, therefore, attempt to develop new political and civic spaces for marginalized groups and communities. This generic idea of rights attempts to link rights of access to and benefits from the development process with the issues of ethnic identity and human dignity, and views the satisfaction of material needs as a pursuit not detached from the spiritual and cultural aspects of human existence (Sheth 2007).

Both the NBSAP and FRA processes critiqued established models of growth within the contemporary Indian state and set out to redefine the meaning of development. In the case of the NBSAP, it built on its critique of national, state, regional and sub-state plans, even as it collaborated with the government. In response to the hegemonies inherent in global politics, the NBSAP aligned with the government in demanding a democratization of the global power structure by exploiting the gaps provided by counter-hegemonic global initiatives, such as the CBD/Convention on Biological Diversity. It used this opportunity to then turn its critique inwards, problematizing the national meaning of development and demanding discussion on the demands of those who have been most directly affected by global economic changes. By redefining issues of development in political and generic terms, the groups working separately on issues such as gender, ecology, human rights or traditional knowledge systems shared the aim of countering hegemonic structures of power at all levels – locally, nationally and globally. The mobilization around the FRA, like that around the NBSAP, also articulated basic development issues in the generic framework of rights. Poverty was viewed no longer as a simple polarization between economic classes but rather as a socio-structural location of exclusion from mainstream development.

> Essentially, a rights-based approach integrates the norms, standards and principles of the international human rights system into the plans, policies and processes of development. A rights-based approach to development includes the following elements: an express linkage to rights, accountability, empowerment, participation, non-discrimination and attention to vulnerable groups.
> (OHCHR 2004)[3]

This framework of rights is a powerful tool to hold public officials more accountable for implementing equitable and effective development policies and progressive legislation. Although this approach is embraced at the level of rhetoric, it does not adequately cover administrative procedures and implementation. Moreover, it is harder to capture the nuances and accountability of non-formal, traditional structures and customary laws within this system. These gaps keep the rights-based approach from moving from formal to substantive equality, though that is its intention. Critics of this approach have also pointed out that the "current body of human rights has been framed from an overwhelmingly European, Rousseauvian perspective of the individual as both the object and subject of rights, but that they have increasingly become the goal and instrument of a modern-day civilizing project in the non-western world" (Batliwala 2010, 1). This became increasingly clear in the historical, colonial experience of formalizing the rights regime in India where community-governed forested landscapes were turned into discrete categories of legal forest and non-forest land (Sivaramakrishnan 2000). In the case of the FRA, the interplay and coexistence of formal rights regimes and customary laws have led to complex instabilities. In contemporary forests, limited state capacity to enforce formal rights regimes and resistance to these has led to the evolution of customary rights regimes, existing in parallel to formal rights. However, customary rights are often in conflict with formal laws leading to tenurial instability and conflicts between the marginalized and the state. Both the NBSAP and the FRA have addressed the issue of customary rights, making specific provisions for community and traditional practices. For instance, within the Forest Rights Act, special provisions have been made for traditional and customary rights such as right of way, collection of soil for household purposes, agricultural concessions and access to religious sites. The larger process of the determination of rights has helped revive community institutions, built social capital and helped establish institutional foundations for management of the community forest resources and the sharing of benefits arising from this protection.

The tensions between rights-based discourse and customary and traditional rights manifest themselves in two ways. First, although the Indian state had instituted both protective and affirmative protection for Scheduled Tribes and other vulnerable sections of society, including several programs such as poverty interventions, watershed programs, forestry plantations and so on, these have often remained at the level of rhetoric and have not adequately impacted the unstable rights regimes in natural resource management. Second, the rights-based discourse often shifts agency to intermediaries and not those who are marginalized. Often these intermediaries (such as lawyers, bureaucrats, NGO leaders and elected representatives) interpret and assert rights on the behalf of local communities but are not entirely accountable to those whose rights they defend. Negotiations that are left to the hands of intermediaries – especially at the level of deliberation with the state, which in spite of their best intentions did occur in both the NBSAP and the FRA processes – often lead to processes created and owned by external champions rather than claim holders. This makes it more difficult to foresee substantive shifts in power at the implementation level.

Democratic mediation

The question of intermediaries in the previous section brings us to another similarity in the two case studies, which is that of mediation. The two case studies share a common feature, which is that of political activism for the inclusion of marginalized groups in decision-making processes, by a third party. Processes that attempt to overcome representational deficits characterize democratic mediation. They help to secure the representation of marginalized groups in formal policy-making spaces and to advance the specific issues at stake. Mediators legitimize advocating for the voice of the marginalized by providing access to forms of knowledge by linking voices to narrow political spaces. Thus, the defining feature of democratic mediators is their belief that "there are groups that are marginalized from formal decision-making who have legitimate interests that are being ignored, and that those groups and their interests ought to be championed" (Piper and Von Lieres 2011, 364).

Activism on behalf of marginalized groups, and not necessarily direct activism, reveals the underlying nature of "democratic mediation." We see in the two case studies that the mediator plays the role of democratic leadership linking ideas from the grassroots to the policy spaces, while advocating for a deeper democracy. Furthermore, it questions the simple divide between participatory spaces framed by the state (invited) and those framed by citizens (invented) in Cornwall's (2002) work by recognizing the important role of democratic activists on "both sides of the equation." Cornwall and Coelho (2005, 1) echo this in their reference to a "participatory sphere" that transcends simple state-society binaries. The important role these mediations play has been highlighted in the case studies, but it also throws up questions of legitimacy and accountability.

While questions of legitimacy are significant to all forms of representation, democratic mediators working within national networks of groups who are marginalized or closely linked to the marginalized are less vulnerable to legitimacy questions. Mediation forms one moment of advocacy in the larger deliberative context. The technical and policy core group in the NBSAP process began as democratic mediators, entrusted with the responsibility of "building bridges" (Apte 2005, 135). The TPCG played an important role in opening up access to public authority and creating spaces for the inclusion of marginalized voices. However, its failure lay in its top-down approach, which was unable to build sustained grassroots mobilizations around user rights in the biodiversity space because of the profile and scope of its mediators. In contrast, what is clear in the mobilization around forest rights is that the bottom-up view of its mediators could make democracy real by its focus on the active struggles of citizens and their organizations. By entering constructive partnerships with formal authorities, democratic mediators ensured that demands were represented in the formal policy process. They made a substantive difference in deliberative politics both by maintaining links with the state and organizing collective action in the public sphere around those demands.

Under what conditions have participatory spaces emerged?

Legitimacy is conferred on civil society organizations based on work and strong linkages to the grassroots level. Policy advocacy is legitimized by evidence of results at the sub-state and regional levels. This often comes from long-term presence in areas, with sufficient experience in engaging stakeholders with the issue at hand. This also allows civil society organizations to wield electoral pressure in the face of conflict with the government over its demands.

The last decade has seen a greater role for civil society organization in the government set-up. The National Advisory Council chaired by Sonia Gandhi saw, for the first time, representatives of social movements and activists. This is the first time in the history of the country where laws were not just pushed but drafted by movements. This interface between the government and civil society has several reasons: (1) there is a subtle difference in the Indian democracy where people are beginning to vote on real issues of governance rather than narrow political agendas or identity politics. This has led to the government's gradual realization that the one way of staying in power is to understand the issues. (2) They also realize that the bureaucratic machinery does not have room for the balanced evaluation of issues because the bureaucracy often tells the politicians in power what they want to hear, keeping political workers alienated from grassroots realities. So in order to feel the pulse of public opinion and ascertain the real needs of the people, they turn to movements. Shekhar Singh supplements this (personal communication, July 2, 2010), referring to policies that are not participatory:

> If it is left up to the government, 90% of the money [from the state] would have never reached the beneficiaries therefore all political advantage would have been lost. People within movements have a sense of ownership and therefore have a stake in the policies which helps governments get reelected. The time for monolithic, single party governments has passed. Now the era is of coalition governments where governments can be voted out on very small margins. It matters now what groups like women's groups or conservation groups think of you. We have come to a certain point in democracy where we are moving from representative democracy to participatory democracy.

There is a clear distinction between movements and NGOs. NGOs are often constructs that can be set up easily and often do not have the legitimacy that movements do. Movements come into existence because a fair number of people share a particular vision. The advantages to linking to movements are that they understand the ground-level concerns and can also tap into a level of talent and commitment, which they otherwise could not have gotten. If the bureaucracy had drafted the legislation, it most likely would have been tilted in favor of and reflected the perspectives of specific narrow interests. These interests tend to be distinct from grassroots reality. The criticism of the deficits in laws that come about with no consultation is overcome by involving public consultation and negotiation at the formulation stage. In addition, as in the case of the Forest Rights Act, movements that are committed to certain issues also support implementation. This adds to the legitimacy of the policy as well as the movement organizations. Movements are ultimately built up after negotiating and addressing core issues, which in the case of forest rights were equity, conservation and long-term sustainability. For movements to be successful there has to be acceptance of leadership and discussions with people in different movements to make space for a broader range of issues. Movements, moreover, have to be a starting point for consensus building, and therefore the creation of forums and spaces are critical to fostering discussion and deliberation.

Political opportunity

In order to capture structural and institutional factors that have an influence on NGO advocacy, political scientists and sociologists have used the analytical tool of political opportunity structure, a term that originated in social movement studies (Kitschelt 1986; Kriesi 2004; Tarrow 1998). Political opportunity structure has been defined as the "consistent – but not necessarily formal or permanent – dimensions of the political environment that provide incentives for collective action by affecting people's expectations for success or failure" (Tarrow 1998, 76–77). It is important to note that structural constraints and incentives work by changing the expectations for success held by activists. Groups engaged in advocacy must perceive particular opportunities and constraints at a specific point in time. In the context of national political systems the most central elements that are considered opportunities or constraints are identified as institutional access, political alignments, influential allies, divided allies, and prospects of facilitation or repression of contentious politics (McAdam et al. 1996, 27; Tarrow 1998, 76–80). More recently, the concept has been expanded to also include "discursive opportunities" (Ferree et al. 2002).

Deliberative systems have distinguished separate spaces in which civil society can engage in advocacy: the empowered space and the public space (Dryzek 2009, 2010; Dryzek and Stevenson 2011). The empowered space is understood as the institutional place dedicated to deliberation and collective decision-making. Institutions that make up the empowered space do not need to be "formally constituted and empowered" (Dryzek 2010, 11). The public space in contrast is "ideally" the realm of "free-ranging and wide-ranging communication, with no barriers limiting who can communicate, and few legal restrictions on what they can say" (Dryzek 2010, 11). The two case studies in this book have had different modes of political opportunity and are compared along three lines of opportunity: (1) access, (2) influential allies and adversaries, and (3) limited repression and facilitation.

Access determines how open or closed a policy space is and how easily groups can enter it and seek responses to their demands. In the case of the NBSAP, it was the ratification of the CBD by the Indian state that signaled a window of opportunity that rights-based advocacy and legislation could be pushed through to a receptive state. When this was supplemented by the state inviting an NGO to spearhead the process of collating a Biodiversity Strategy Action Plan, it seemed like the opening up of an empowered space. The actual source of empowerment however is not only the normative power of institutions like the Convention on Biological Diversity that gives credence to rights-based claims. Rather, it is the hierarchical structure of biodiversity negotiations in the international sphere. The TPCG and the NGO Kalpavriksh, chosen as the coordinator for the process, explicitly chose this mode of participatory engagement because the Indian state was engaged in conflictual negotiations at the international level. Human rights frameworks and international institutions provide opportunities to marginalized groups in terms of claims making, but ultimately national interests and asymmetries of power dictate the acts of states. Thus, the claim making norms of international institutions have leverage only within the specific political opportunity accorded by the imbalance of power at the international level. In contrast, as explained in the case study on the Forest Rights Act, the political opportunity seized by activists demanding rights-based legislation for forest-dependent communities came with the shift in political power with the election victory of a government backed by the Left Front, who were more conducive to the demands of the activists. In addition, the threat of violent uprising amongst forest-dependent communities in poor states was turning into a serious internal security threat. Many commentators have pointed out that legislations like the FRA are clearly populist, with political goals in mind. However, activists were able to exploit political identities through the creation of cohesive collective demands in the public space. The opportunity that was exploited was the opening up of the institutional space to these demands.

The second aspect of political opportunity is the role of influential allies and adversaries that can shift the balance of power for groups engaged in advocacy. The role of both allies and adversaries has been captured in this interview with a member of the steering committee for the NBSAP:

> The NBSAP came to the IIPA [Indian Institute of Public Administration] initially. Ashish [founder, Kalpavriksh] used to be a research associate. The special secretary in the ministry had done his course in the IIPA and knew Ashish well and had great regard and respect for Ashish and so he picked him. Then [during the process] there was a change of bureaucratic leadership and the man who came [into the MoEF] was anti-NGO which led to personality clashes. There were also many recommendations in the NBSAP which many governments would have not accepted but the regime at that point was too arrogant to even talk to. Thus more than a weakness in the process it was a clash of personality. Ordinarily it was unusual to give the process over to a 'radical' NGO like Kalpavriksh who has traditionally been opposed to the government. The right thing for the government would have been to sit and have a give and take with Kalpavriksh and have a dialogue and discussion
>
> (member of steering committee, personal communication, July 2, 2010)

This is supplemented by Ashish Kothari, who confirmed that one of the crucial factors for the change in the balance of power was that 2003–06 saw some regressive changes being made to the ministry under the then secretary of forests who took an intense dislike to an NGO-led process. This became a major constraint in garnering support for the plan within the MoEF, and as a result it could not be accepted as national policy. At this juncture the process of building up support for the NBSAP had to be taken out of the empowered space to the public space in order to put pressure on the government through advocacy.

In contrast, the Forest Rights Act had powerful allies in the Communist Party of India, who lobbied against its dilution. They had influence with the congress, their coalition party at the central level, to get the act enacted. In addition, they supported mass rallies and disseminated information in order to mobilize the grassroots to put pressure on the government to enact the law. CPI(M) manifestos in states like the Himachal explicitly referred to the act as landmark, pro-poor legislation championed by the party, stating: "The Left, particularly the CPI(M), has played a pivotal role in the struggle for equality of the Dalits and tribals. Of the total land that has been distributed in the country, the Left ruled states account for more than 70%" (Panwar 2009, 1). Though this clearly ties in to the commentaries on the FRA pushing populist legislation, what is clear is that political patronage played an important role in pushing this legislation through and was crucial in the deliberative negotiation within the empowered space.

The third aspect of limited repression and facilitation refers to the activities of states or governmental agencies to restrain or foster the political participation of civil society, which is explained in greater detail in the case studies. Domestic advocacy coalitions can use global instruments to leverage their position when they take the ratification of treaties as their point of departure. However, this has to be supplemented with a responsive state, political strategizing and the frames most conducive to policy adoption. What is clear through the case studies are that India has a strong (centralized) state facing strong social organizations. Although access of transnational actors might be easier than in a state-dominated case, the policy impact could be limited due to structural problems of societal and political institutions that may deter policy change. Successful policy change can arise from domestic processes and actors, altering domestic structure during the process. In addition, the state will only facilitate civil society organizations if they are backed by a considerable grassroots presence.

What roles do narratives play?

Policy narratives are the stories that people tell to make sense of policy issues and underpin policy solutions (Roe 1994). They contain important structural features, including a coherent account of events and a beginning, middle and end. They focus on agency and contain a normative message or moral to be decoded by the audience (Stone 2002). A leading group of theorists in public policy and governance indicate that such policy narratives are an important aspect of political advocacy (Kaplan 1986; Roe 1994; Hajer 1995; Stone 2002; Fischer 2003). Moreover, a growing body of empirical literature has demonstrated the impact that these narratives have in applied policy settings (Bedsworth et al. 2004; McBeth et al. 2010). Fisher (1987) supplements this, pointing out that people always employ narratives when thinking and talking about complex social issues. Likewise, policy narrative literature empirically proves the prevalence of policy narratives in debates held in the media and between powerful interest blocs (McBeth et al. 2010).

Deliberation is regarded as a communicative, talk-centric activity; however, the specifics of how communication actually occurs are subject to much disagreement. One definition put forward by Gutmann and Thompson (1996) is the idea of "mutual justification," whereby political actors must communicate justification for adopting particular plans and strategies to others with an interest in the same issue. This idea of justification links to the persuasive element in advocacy strategies within deliberation where policy narratives are used to persuade other actors to follow particular policy prescriptions. In the context of the two case studies, we see policy narratives legitimize and justify certain positions that actors take within a political subsystem. The "historic marginalization" of indigenous people and the "tiger versus tribal" tropes all allude to specific historically constructed narratives that strengthen and justify actors' positions in deliberative forums. Thus, policy narratives do important sense-making work, simplifying complex issues and providing people with frameworks through which to understand their experiences and observations, rationalize public debate about the possible courses of action and buttress political actors' sense of identity (Hajer 1995).

The dramatic element of policy narratives that is underscored by Stone (2002) also plays an important role in contentious policy deliberation. In order to justify a particular course of action and give it legitimacy, a dramatic story arc must be followed. Stone shows that most stories fall within one of two genres – narratives of decline, and narratives of control. In the former, the narrative begins with a pessimistic description of the status quo, before outlining how the right policy can rescue the situation from catastrophe. This is easily captured by the Forest Rights Movement, which painted a picture of bleak, historical dispossession suggesting that armed civil struggle would be the only option left for indigenous people to claim their rights. In the latter, the narrative begins with a status quo of chaos, suggesting that the right policy can bring calm and order back to the policy setting. This is captured in the narratives of equity tied to biodiversity, where a picture of a country in the clutches of rampant bio-piracy, losing its vast wealth of indigenous knowledge because of short-sighted and exclusionist policies, was painted. In both cases, narrators are at pains to depict matters at extremes, emphasizing the drama and emotion in order to delegitimize alternative approaches to equity, conservation and sustainability. Stone also points out that policy narratives typically have a regular "cast of characters," with actors in the key roles of heroes, villains and victims, which is also clear in the case study on forest rights that framed debates between tigers (and forest bureaucracy) against the tribal. Heroes and villains are readily identified through the evolution of the issues of biodiversity and forest rights. Women, peasants, Adivasis, forest dwellers and fisher folk – the marginalized and the dispossessed – are pitted against the forces of big business, (Western) reductionist science, corporate interests and a neo-liberal government.

These narratives of postcolonial environmental decay and possible restoration have been subject to careful critical review both within India and beyond (Williams and Mawdsley 2006). One has to resist the portrayal of the state as a monolithic entity, completely captured by external interest, and the assumption that "ecosystem people" want to live a simple life that rejects the evils of materialistic consumerism (for alternative views, see Rangan 2000; Osella and Osella 2000). However, it is important to note the efficacy of these populist narratives in garnering social support and mobilizing people around a number of political issues facing India.

The Technical and Policy Core Group in the making of the National Biodiversity Strategy and Action Plan (NBSAP) saw political marginalization and the lack of rights over resources as the underlying cause of both biodiversity loss and poverty. Their motive in creating large-scale participation around the issue of biodiversity was to ensure that political decision- making about biodiversity included the differing values of biodiversity. They wanted to engage with the different values of biodiversity and expose which narratives take precedence. This was to ensure that the voice and influence of those who are marginalized from key decision-making processes were strengthened, and that those whose voices were traditionally marginalized were given the institutional space to negotiate where interests diverged.

The TPCG understood marginalization as those who were sidelined from "official environmental planning, and not to marginalization in general (i.e. someone who occupies the mainstream in civic life may nevertheless be marginalized from planning)" (Apte 2005, 22).

Using Arnstein's ladder (Table 5.1), the TPCG clarifies that most government-led policy processes stop at tokenism where citizens may hear and be heard but where they do not have the influence to make sure that their views will impact policy. They underline that "placation is a higher level of tokenism because the ground-rules allow have-nots to advise, but retain for the power-holders the continued right to decide" (Apte 2005, 22). In the case of the NBSAP, the TPCG by framing their concerns in terms of political marginalization and linking this to national threats of bio-piracy acquired a legitimate voice and platform to air grievances, suggest reform and put pressure on national governments.

The civil society actors around the Forest Rights Act also utilized the frame of *marginalization*, but it saw it as a result of centuries of dispossession from traditional lands and pushed for a national legislation that corrected a historic injustice. Like the NBSAP, these CSOs also underlined the importance of political marginalization of indigenous people, not just in contesting elections because of the lack of financial resources but also in representation in the executive wings of government. It was pointed out that even at the level of local government (*panchayat*) in areas with tribal concentrations, non-tribals manipulated the constituencies. Indigenous communities were also seen to be having only "token" representation

Table 5.1 Degrees of marginalization (source: Arnstein 1969)

Manipulation	Therapy	Informing	Consulta-tion	Placation	Partner-ship	Delegated power	Citizen control
Non-participation		Degrees of tokenism			Degrees of citizen power		

in the cabinet rather than in terms of their numerical strength, which was largely guided by constitutional compulsions (Planning Commission 2008).

One of the main objectives of NBSAP and FRA processes was to promote equity in communities, while adapting to traditional management systems of bio-diversity and forests. It aimed to ensure access of all users to resources to meet basic needs, to promote participation in decision-making relating to resources they depend on and to ensure equitable benefit-sharing mechanisms. In doing so the ultimate aim was both to adhere to a global rights-based discourse and to promote sustainable use of resources to ensure inter-generational equity. This was imperative as the government was unable to provide access for the poor and mar-ginalized to decision-making forums, forest resources and community funds. The elites in rural communities carved out very influential spaces in decision-making, and this led to further marginalization of communities and Adivasis. In the case of forests, as demonstrated by the case studies, many restrictions on access and use were historical in nature, managed by the Department of Forests and donor agen-cies at the national level and by powerful, rural elites at the community level. This resulted in a consolidation of power in the processes of planning and implement-ing programs (Ojha and Timsina 2008; Timsina 2002).

In response, the NBSAP process attempted to tackle these issues with an inclu-sive and interactive model which facilitated critical understanding of biodiversity collectively, aiming for holistic intervention in partnership with the government, challenging long-standing power relations and positively affecting equity. As noted by Tejaswini Apte (2006, 6):

> The NBSAP planning strategy consistently emphasized that the process of putting the plan together was as important as the final product. In other words, regardless of what might come out of the final plan, the process itself was meant to increase awareness of biodiversity, empower people through participation, inspire local initiatives to begin implementation of local plans, and so on. In this sense, the NBSAP process became a form of activism.

This is relevant for the theory of deliberation that puts more emphasis on the process rather than the outcomes as well as the emphasis on the process of collect-ing and discussing different approaches, interpretations and inter-subjective under-standings of the resource of biodiversity. This focus on implementing local plans and initiatives was divorced from the more procedural aspects of policy-making at the central state. It was envisioned as a pathway to guard local and commu-nity interpretations and traditions around biodiversity against the homogenizing resource policies at the state level. Thus, advocacy at the local level played a cru-cial role in guarding local-level processes against state capture.

In a departure from the usual political process, using opportunities provided by the norms of the CBD, the NBSAP process provided the space for unconstrained dialogues at the national, state and sub-state level. Though the government is not obliged to involve civil society organizations in any policy formulation, windows of opportunity encompassed by the CBD directives were utilized to bring the issues of access and control of the marginalized to the forefront. The TPCG and its networks of civil society organizations attempted to facilitate the negotiation needed to achieve equity outcomes through regular and critically reflective interactions among different sub-groups within the policy sub-system. This process focused on governance and empowerment, and they achieved this by creating forums (in the form of national and regional workshops) and mass participation (biodiversity festivals, public hearings) that brought diverse perspectives, interests, knowledge and information from within and beyond the community into the discussion, highlighting the knowledge gaps in usual policy processes.

The NBSAP process also identified several grey areas in their understanding of the marginalized. Focusing specifically on CSOs and the state, Apte (2005) points out that all organizations cannot be unambiguously termed politically marginalized. Many CSOs are influential in the public space and routinely consulted on policy issues. In contrast, there are several other organizations that remain firmly outside the public space and many grey areas in between. What is clear in the NBSAP process is that "participation" in the end did not go beyond the creation of information and consultation, falling short of a true partnership between the state and civil society. This suggests a type of staged democracy, where people's inputs had very little influence in the final draft of the NBSAP. This adheres to Clark's (1995) opinion that civil society's contribution in the development sector and the process of social change largely depend on their relationship with state authorities and that the state's treatment of them can be anywhere between benevolent to hostile.

In contrast, the bottom-up process of mobilizing around the Forest Rights Act involved social movement organizations, community-based organizations and self-help groups. They facilitated the mobilization of community resources, the development of forest operational plans and the development of local-level enterprises and infrastructure. Civil society was also actively involved in disseminating the frame of historical dispossession and actively advocated for political space for those who had been traditionally marginalized in society and politics. They demanded that representatives at the national level had to contribute to the policy formulation processes. Civil and social movement organizations also played an important role in long-term identification of the marginalized and identity creation through creating institutional structures like the forum of forest workers. In addition, over several years there was an institutional democratization of organizational processes through the creation of the community forest management (CFM) network, *Van Suraksha Samitis* (VSS, forest protection committees), Forest Rights Committees and *gram sabhas* (village councils) at many different levels. These various organizational innovations, however, have come with their own problems. The government and forest bureaucracy have given precedence to the principles of the VSS, which were instituted under the joint forest management schemes, rather than forest rights committees, which came about through the Forest Rights Act.

In the course of mobilizing people around forest rights, civil society organizations conducted several awareness-building exercises on rights that people should be entitled to. They also began to establish good working relationships with key people in moderate conservation groups and government institutions to foster the exchange of knowledge, contribute in the policy-making process and at the same time communicate the complexity of socio-political and cultural issues, including equity. One of the most contentious issues was the inclusion of the non-tribal in the form of "Other Forest Dwellers" in the FRA. The pro-tribal lobby pushed for the inclusion of the category of "Other Forest Dwellers," who, though not strictly living in forests, were dependent on forest resources. As they were not historically marginalized, the government did not accord them any particular disadvantaged status or its consequent benefits.

Other Forest Dwellers are understood by the Forest Rights Act as "any member or community who has for at least three generations prior to the 13th day of December, 2005 primarily resided in and who depend on the forest or forests land for bona fide livelihood needs" (FRA 2006, 1 sec. 2 [o]). Critics have pointed out that this definition remains ambiguous as the requirement to "reside in . . . forests or forest land" (ibid.) is applicable to both tribal communities and non-tribal communities. Most forest dwellers, unlike the tribals, do not live in areas recorded as forest land but rather live in revenue land, which often falls on the fringes of forests on whose resources they are dependent (*Economic and Political Weekly* 2007).

This extension of rights was a bone of contention for conservationists who pointed out that there was very much ambiguity surrounding the criteria for determining the beneficiaries of the act. This could have an adverse impact on conservation activities as powerful lobbies could use the act to appropriate tracts of forests. They also pointed out that other forest dwellers, unlike the tribal communities, often have no strong traditional and cultural links to the forest. Sahu (2010, pt. 1) notes:

> The most dangerous misunderstanding among the people is the aspect of 'non-tribals' having to prove 75 years of existence on the land before they can claim rights over the same. In many places, it has been propagated that the Act does not allow non-tribals to make either individual or community claims. Added to this is the typical feudal setting that still exists in rural India in which non-tribal landlords and their cronies virtually rule over tribals and other weaker sections of society. NGOs have not made enough effort to address the new 'class' issues emerging in various parts of the state, in the wake of FRA 2006.

Thus, we find that both processes saw civil society actors playing a key role in discourses of critical awareness leading to mass participation, advocacy and capacity building and highlighting justice and equity issues. Resurrecting the narratives of the marginalized is not unproblematic. CSOs must be self-reflexive about the ambiguities involved in demarcating the marginalized, even though the frame is useful for addressing justice and equity issues.

The importance of global norms

With regard to two of India's biodiversity policies, the National Biological Diversity Act (2002) and the NBSAP (2009), it has been shown that the domestic coalitions did draw strength from the ratification of the CBD. This provided a window of opportunity that rights-based legislation could be pushed through to a receptive state. These examples suggest that first by ratifying global conventions, relational forces within civil society will be altered, strengthening the position of certain groups, like the rights-centered advocacy groups and discourses, and sidelining others, like the conservationists. Second, there could be an opportunity that advocacy organizations could restructure political processes and planning vis-à-vis the state. Conventions like the CBD can kick-start rights-based advocacy even in settings where these approaches have very little resonance with the state. The CBD outlined broad ideals and norms that were open to interpretation within a domestic policy space. This study clearly demonstrates that the CBD gave credence to the actors involved in biodiversity policy formulation in India. It gave them an overarching frame of reference that could be adapted to the domestic context. The ideas of mass participation, mechanisms of access and benefit sharing and a rights-based approach were absorbed by domestic actors in their advocacy and their efforts to open up the policy process to a more diverse set of voices.

In the case of forest rights, the role of global norms and frameworks had a more subtle position where they became symbols of opposition, where advocacy had to effectively counter hegemonic structures of power, both nationally and globally. This translated to "strengthening campaigns against the commercialisation of the forests and oppose G-20, European Union, WTO, ASEAN and other international bodies pushing capitalist agenda globally"[4] and building alliances with counter-hegemonic global initiatives and movements.

There was a process of reformulating and narrative building around old issues of development recast in new political terms, although their objective remains unchanged – for instance, that those sections of society, which have been historically marginalized, get their due as producers in the economy and are given voice and access linked to substantive citizenship. Accordingly, development was considered a political struggle by which to achieve participation of civil society in defining goals of inclusive development and articulating strategies. In pitting this vision against global hegemonic structures, they define development as a non-hegemonic, pluralistic process, in the articulation of which they use insights and criteria evolved through their own struggles. In this process they increasingly drew strength from globally debated issues such as feminism, ecology and human rights and embedded these debates in the social, economic and cultural specificities of India. Consequently, the idea was to make development a bottom-up process, and participation was used to capture the ideas of the grassroots and infuse them into governance structures and thereby challenge the power relations on which the conventional model of development rests.

What does this imply for the practice and theory of deliberative democracy?

The two case studies provide practical insights for civil society organizations' role in deliberative democracy. First, it demonstrates that supporting deliberation requires long-term strategy and practices that support critical thinking and political agency. This is to ensure that different viewpoints find spaces to be articulated and that these are then reflected in a broader political context. Second, these cases suggest important lessons on how to approach nascent spaces for deliberation, when collaboration is necessary and when pressure tactics or collective action is required.

This has implications for theory as advocacy plays a critical role in linking citizens and claim making to democratic governance. By collectively putting together meanings and knowledge based on local experience and perspectives, organizations develop public claims that are demanded from the government or broader society (Stivers 2002). The case studies also support the argument that civil society organizations play a dual role in governance: first, by developing new discourses and, second, by promoting them in other spheres to challenge the state (Habermas 1984; Fraser 1997). Civil society organizations acting as mediators between the state and a broader civil society do not just develop new discursive horizons (Dryzek 2000) but also develop alternative spaces to the ones offered by state-led deliberation. These alternative models of deliberation encompass both the spatial aspect (that is, where public deliberation takes place) and the communicative aspect (how deliberation is structured).

Within the spatial action, deliberative democracy underlines the importance of forums and the interactions of those forums with other spheres in the deliberative system (Mansbridge 1999). The idealized, rational notion of the public sphere has been criticized for its idealism. Fraser (1990), Mouffe (1996), and Young (1990) have pointed out that in its idealized form, deliberation has not paid attention to coercive forms of power in public discourse, has excluded affective modes of communication in favor of rational discourse, and has tended to promote consensus as the purpose of deliberation. Other approaches have attempted to address these gaps. For instance, Fraser (1990) argued that strong publics, whose discourse encompasses both opinion-formation and decision-making, should be differentiated from weak publics, whose deliberative practice consists exclusively of opinion-formation and does not encompass decision-making. Habermas (2005, 388) supplemented this by contrasting two types of political deliberation: "(a) among citizens within the informal public sphere and (b) among politicians or representatives within formal settings." The radicalization of the public sphere and the differentiation between strong and weak publics are particularly useful for understanding the two processes explained in the case studies as they encompass both strong and weak publics in planning processes. The role of representation thus becomes even more valuable among deliberations within formal settings, because for any real impact on policy, this has to be supplemented by opinion formation at alternative forums or informal public spheres like the media. This shows that advocacy is an effective way to bridge the strong and weak public sphere as strategies that are created and supported in the public sphere have to effectively impact representatives in formal settings.

The idea of public action emphasizes the role of social organizations in organizing the public to advocate for social change, rather than fill gaps that the government or public has overlooked. A discursive perspective suggests that this important function involves synthesizing grassroots or local knowledge and integrating these perspectives into conversation about public issues. This also emphasizes the role of CSOs in creating more flexible forms of political engagement (Fischer 2006). This flexibility in communication, particularly in reference to the importance of local knowledge, is expanded by the use of clusters of communication tools to elicit mass participation. These communication strategies are tailored to the local and grassroots level to glean the maximum knowledge and insight that could then be translated to demands at a more formal deliberative setting. A subject's personal relationships, interpretations and understandings of biodiversity or forest rights are given emphasis. With their emphasis on decentralized participation, CSOs were able to link voice and empowerment to substantive rights of communities, including, for example, the rights to food and development, along with customary rights to land and water. To this end, CSOs mobilized powerful narratives against the entrenched ideas of ecological modernization that were being propagated by the state to counter the elitism of the traditional law-making factions.

Public policy debates argue that participation is intrinsic to modern democratic politics, promoting government accountability and linking them to citizens' needs and preferences. For some, it implies a deepening of democratic deliberation, while for others it represents grassroots resistance and mobilization against powerful political elites and government agendas of neo-liberalization. The case studies illustrate that both conflict and consensus form part of the policy formulation process. As we saw in the successful legislative outcome for the Forest Rights Act, bottom-up mobilization must complement deliberation in more structured dialogue with the state in order to keep the state from retreating and create legitimacy for the demands that civil society is making of it.

While these questions around consensus or agreement are central to democratic theory, there is a polarization between those seeking consensus through deliberation and rational decision-making and those who underline political contestation as democracy's central feature. I argue that the ensuing conflict can be traced in the top-down and bottom-up processes of deliberation. The distinction between top-down and bottom-up processes is not mutually exclusive. In fact, they overlap to form a "hybrid" with differing strategies of influence by civil society being experimented with at different levels. This idea of a "cycle of contestation and consensus" complements the notion of "social movement cycles" (e.g. Tarrow 1994) and resembles the "agonistic" model that is open to multiple voices and the informal and unstructured nature of public discussion. Civil society is called upon to play an informal role both outside and in opposition to the state, which require both communicative and strategic behavior (Hendriks 2002). In this set-up, groups do not merely seek to persuade each other but want to influence the court of public opinion. This is done by forcefully asserting the narratives, issues, values and interpretations that they believe should be binding for everybody, sometimes even through "sensational actions, mass protests, and incessant campaigning" (Habermas 1996, 381). The most important normative requirement for this kind of agonistic narrative is that of authenticity. The state may not recognize these narratives as factually true or morally appropriate but have to be convinced that there is an authentic commitment to those values that the public mobilizes around. The state is then forced to react, collaborate, expand its formal spaces to popular representatives and finally allow for policy change. Even though there are periods when compliance and compromise are used in democracies to achieve goals, conflict remains institutionalized and gets normalized in political processes. The conflict in the plurality of opinions and the push-pull of democratic policies is in fact what keeps the state under scrutiny and therefore accountable and civil society engaged.

It is clear that grassroots mobilizations take place within a wide range of spaces and that those involved in representational advocacy make use of different strategies in presenting demands and pushing for results. Informal processes like mobilization do not follow any specific path but remain situated in specific contexts. In these case studies we see a variety of arenas where issues are negotiated. These include formal channels like the courts and parliament and forums that range from the ones provided by the central government to the state governments, as well as the sub-states and regional levels. Informal forms of eliciting participation and deliberation are utilized, rooted in culturally specific modes of communicating like public hearings, festivals and workshops. Strategies rely on directly oppositional forms (e.g. rallies, demonstrations and picketing) to state-led partnerships like collating biodiversity registers. In addressing conflict, strategies focus on mechanisms of survival or coping strategies, which at other times turn to resistance, challenging structural ideas. Thus, there can be no rigid conceptual barrier between invented and invited spaces, as they share a number of strategies. The spaces for practicing citizenship are not mutually exclusive. Collective action moves between these different points, using a varied set of tools and spaces to supplement deliberation in formal spaces. What distinguishes formal deliberative spaces and the spaces of mobilization are not the affiliated groups who may move between the two and occupy both spaces. The distinction is that actions taken within invited spaces of deliberation, no matter how innovative, are sanctioned by government and donors and are therefore restrained by funds and the discretion of the government.

Within invented spaces, mobilization thrusts demands to the invited spaces and is characterized by defiance against the prevailing status quo, although arriving at a consensus can only happen in a deliberative process within state-sanctioned deliberative arenas. One such space is the Joint Parliamentary Committee (JPC), which is an ad hoc parliamentary committee constituted by the Indian Parliament. Mandated to inquire into any prominent subject concerning the country, the JPC is constituted either through a motion adopted by one house and concurred by the other or through communication between the presiding officers of the two houses. The members are either elected by the houses or nominated by the presiding officers. As in the case of other parliamentary committees, they are drawn from different groups. The strength of a JPC may vary. In the case of the forest rights bill, it had twenty-eight members from the Lok Sabha (Lower House) and Rajya Sabha (Upper House of Parliament). The JPC is required to conduct a detailed enquiry in the matter and give its findings in the form of a report. In order to do this, it can collect oral and written evidence, inspect all documents related to the subject and summon anybody to appear before it except for ministers and the prime minister. Members who do not agree to the recommendations of the committee may write dissent notes; these form a part of the JPC records. For the forest rights bill, in addition to members of the JPC, 109 associations, organizations and individuals presented their memoranda to the joint committee from across the country. In the face of strong conflict surrounding the act, the JPC was the purely deliberative face of the state and was crucial in consensus forming and conferring legitimacy on advocate's positions.

This study puts forth the proposition that successful policies must be negotiated in formal state-sanctioned spaces. However, mobilization around certain issues is a key ingredient as it confers legitimacy on democratic mediators. They can call upon mobilization at critical junctures of conflict during the policy process. Thus, grassroots mobilization practices invent new spaces of citizenship practice and offer a significant push to achieving substantive citizenship. It helps to "expand the public sphere" (Rose 2000, 18) to more equitable policies. This book sharpens the range of collective action that is engaged with at the level of the marginalized, through both invented and invited spaces of participation, and points out under what conditions successful deliberation is achieved. It provides a good opportunity to examine empirically the plausibility of the deliberative democracy's institutional blueprint.

Strong political support is often necessary for truly transformative deliberative empowerment. Yet, it is clear that the state is not a benevolent or politically neutral character. Instead it has clear preferences, although it does not clearly spell out what it wants. In the case studies examined, the overarching narrative of ecological modernization proposed by the state is made clear through the processes, and this is part of the global neoliberal discourses appropriated by the state. This means the Indian state also selectively chooses agendas to give support to and remains ambiguous about others. Civil society perceives the state as strategically pushing the interests of the strong and neglecting the demands of the weak. The evidence presented here has important implications that, I would argue, go well beyond the specific empirical material examined. In particular, these cases highlight that participation within the limits of specific institutional collaboration has lost much of its transformative connotation and potential. Instead participation without the element of advocacy could contribute to the explanation of the apparent ease with which the notion of societal participation has been incorporated into the mainstream policy discourses of several international organizations (like the World Bank, among others) and several national and regional policy-making organizations.

According to Dryzek (2000), this absorption of oppositional processes into the state is sometimes a gain and sometimes a loss. He points out that once these oppositional forces are absorbed by the state, the state has less to fear by way of public protests. This, according to him, is a democratic loss in terms of a "less discursively vital civil society" (109). However, the neat cleavages between cooptation and democratic loss have to be problematized by adding the dimension of advocacy to the policy process. I argue that if advocacy is maintained throughout the policy process, the legitimacy of private spheres for formal deliberation is maintained by connecting it to social spaces. These exist at different levels outside formal official spaces, such as those created for public consultation or people's forums. Communities and individuals bring into these spaces experiences from their everyday lives and their experiences in other spaces. Advocates who straddle both official and unofficial spaces, in particular, have to then orient themselves to these spaces and decide how to employ the ideas discussed in them. This is crucial in conferring legitimacy on both official and unofficial spaces. It marks the difference between a policy that suffers democratic losses after mobilization has been absorbed into the state and a successful policy where the arena for deliberation has been expanded through mobilization, which confers legitimacy and democratic control on representatives collaborating with public actors in the formal spaces of deliberation. The official or claimed space then becomes the nucleus of advocacy. My point of departure from Dryzek's (2000) analysis is in relation to his understanding that once oppositional forces have been absorbed by the state, it has much less to fear in terms of protest. What both case studies show is that inclusion into state processes is merely the starting point of political advocacy. These case studies point to strategies used, including direct action and civil disobedience, to keep issues alive in the public sphere even after they have been coopted into private/formal spaces within the institutional machinery of the state.

There was also a crucial role played by democratic mediators who encouraged the setting up of disaggregated deliberative spaces. They infused a deliberative culture within those spaces and laid the groundwork for advocacy to be continued throughout the process. Mediators played a leadership role linking ideas from the grassroots to the policy spaces, while advocating for a deeper democracy and conveying the failures and successes of the formal process back to the informal deliberative spaces that could then be discussed, negotiated or resisted. As a result, those moving within the informal spaces were infused with the message that their opinions were being communicated back to leadership and that their narratives of resistance, local knowledge and experiences were being highlighted at a national stage. It was also recognized that participatory practices would have to be infused into beneficiary or user groups surrounding the policies before they could recognize themselves as a community. This was further complicated by the breadth and diversity of a country like India where the same resources (e.g. forests) had very different communities and interpretations surrounding them. It was crucial to have both community leaders who would assist the process and experts and activists who could assist the communities in coordinating local struggles into scales that would matter to political leadership.

One lesson to be drawn from this is that formal and informal spaces have to be linked to create political ownership at the central level. It is clear from the case studies that "claimed" processes, ones evolving from civil society, are more successful than state-led processes that limit the transformative potential of informal deliberative spheres. Civil society on its own needs pathways to transfer its ideas into formal deliberative forums within the state in order to allow for truly transformative policy-making. To this end, democratic mediators are crucial as we see in the process of the FRA. They are needed for two activities: first, they help recognize and take advantage of political opportunities at specific points of time. This can be done only by advocates who have the critical and analytical acumen to decide on when issues can be positioned in a national stage and how to utilize those opportunities for the benefit of the more marginalized factions of society. Second, democratic mediators are necessary for countering and under-cutting hegemonic narratives in nuanced ways and representing them in ways that will allow for collective mobilization at a national stage. Because deliberation remains an activity that is given a lot of political currency by elites, advocacy has to be aligned to deliberative processes in order to keep it highly visible in the public sphere.

Theories of deliberative democracy rarely mention political activities such as activism or advocacy (Young 2001). A criticism that supporters of deliberative democracy may hold towards activism or advocacy, especially in the form of advocacy on behalf of a certain group, is that advocates tend to orient themselves towards identity-based or group politics rather than a commitment to universal principles. Pressure group interest-based politics allows people to organize into groups promoting specific interests. They pressure or convince policymakers to serve those interests by means of lobbying, contributions to political parties or mobilizing votes for or against particular political parties. They feel very little obligation to engage with points of views that are in opposition to their own interests or to resolve conflicts through dialogue. Their aim is to organize around a particular narrow interest or identity and engage in power politics to secure the most victories for their interests. The case studies demonstrate that group interests in my case studies are carefully constructed and linked to substantive and universal rights. The main constellations of actors in this process are not just interested in narrow identity-based groups. They may make a concerted effort to engage with different factions of society and their ideas of what biodiversity or forest rights ought to be. Both the processes were set up around rights-oriented discourses and historical marginalization, critiquing established models of growth within the contemporary Indian state and setting out to redefine the meaning of development. Rather than furthering solely strategic interests, both processes positioned themselves as means of fighting for universal justice and critiquing the social, economic and political institutions that were producing unjust structural inequalities. Activism was seen as pre-emptive to democratic loss and cooptation and inherently useful in expressing dissent in the public sphere. This was especially the case once the issue entered the formal deliberative space and it was demonstrated that conferring legitimacy on existing institutions and participating in the structuring of those institutions could lead to compromises that could silence dissent.

Theorists before have discussed choices between state and civil society. Claus Offe (1990, 243; cited in Dryzek 2000, 108) for example discusses three phases that movements go through: the informal and militant "take off" phase; the "stagnation" or "consolidation" phase, where leaders, organizations and membership is defined; and the "institutionalization" phase, where entry is gained into the state and there is access to real political power. This has also been termed as cooption

(Dryzek 2000). Others (Rucht 1990; Cohen and Arato 1992) point out that movements and collective action do not follow simple formulas and have many more choices at their disposal, including sustained action in both state and civil society. Some have proposed dualistic strategies (Wainwright 1994; Cohen and Arato 1992). In these strategies, it has been pointed out that movements often diverge with communicative, discursive politics. They target political and economic institutions that include rational politics of inclusion and reform, while the movement in civil society continues to reinterpret norms and identities and focus on developing more egalitarian politics (Cohen and Arato 1992). However, this has to be supported by independent relationships and access to political power. In this scenario, extra-parliamentary activity can provide continued support to parliamentary or legislative processes rather than substitute for them. Within deliberation, advocacy has been largely ignored with few studies (Young 2001; D'Arcy 2007) that look at the emerging activities of CSOs or individuals in deliberative spaces. Advocacy in this study is clearly linked to the opening up for formal deliberative structures and linked to them by micro-forums in civil society. Thus, activism in the public sphere is clearly linked to political opportunities and deliberative victories in the formal space within the state. In that sense it follows a dualistic strategy, which is aligned to the prescriptive idea that "only if there is a continuation of politics by extra parliamentary means will democracy be able to establish limits to the power of the dominant class" (Fisk 1989, 178–79).

The conclusion I draw from linking advocacy to deliberative democracy is that it is a barrier against the democratic loss, which may follow the assimilation of policies into the state or private sphere. Linking advocacy to deliberation on policies may make formal spaces more legitimate and accountable to the people whose issues are being discussed by representatives. It may also help to avoid assimilation and symbolic rewards. In addition, continued advocacy in the public sphere may allow for discursive contestation that hones civil society's inter-subjective understandings. Through discursive contestation, actors are given time and space to confront and discuss different interpretations and points of view and to try to achieve a rationally grounded consensus. My claim is that non-deliberative means of advocacy only legitimize the decisions made in strong public spheres like the legislature or parliament. In order to negotiate within an institutional deliberative space and arrive at a more participatory policy process, civil society actors must engage in extra-parliamentary forms of strategic advocacy in the public sphere to sustain those spaces of deliberation, legitimize them and guard them against dominant political interests and discourses.

Broader implications

While it must be noted that this book meditates on formulation, the scenario of implementation of environmental regulation has followed a declining curve. In terms of strategies, one finds innovative movements to counter the lack of transparency in environmental implementation and formulation. For instance in 2005, CSOs conducted a "funeral" of the Ministry of Environment and Forests (MoEF) and symbolically issued a death certificate, under the banner of "MoEF Chalo," to overturn the proposed reform of the Environment Impact Assessment Notification and the draft National Environmental Policy, formulated without consultation with elected representatives and the wide public. Consultation has been the rally cry for activists and citizens over the last decade. In response in 2013, the Congress National Advisory Council recommended mandatory pre-legislative consultation be introduced for all proposed laws; it said that this would take India from a "representative democracy to a participatory, deliberative democracy" (*Hindu* 2014). This was done to involve as many stakeholders while finalizing draft laws and amendments to existing laws. In particular, institutionalizing the idea of a deliberative democracy is particularly significant as it filters into the mainstream political lexicon. Government communication also highlighted the necessity for laws to be communicated to the public in simple language, allowing for a more inclusive interpretation of the laws in question. However, this being an executive order means that there are limits in terms of the scale at which it is enforceable by law.

The Aam Aadmi Party (AAP; the common man's party), who fought and won the Delhi elections in 2015, further highlighted deliberative and participatory governance. Their core belief is in the superiority of decentralized, participatory democracy over top-down administration. AAP leader Arvind Kejriwal's 2012 book *Swaraj* spells out his detailed plan for the country that primarily revolves around the ability of local administrative units (*gram sabhas* [village committees] and *mohallah sabhas* [neighborhood committees]) to be more responsive to the choices of local people. At the same time, nascent spaces of participation, allowing access and voice, are constantly threatened with marginalization. Post-2014 election India is seeing a conflict of visions at different scales. Much like the policies discussed, there are contesting visions of what democracy will look like in an emerging India. While one thread supports decentralization, participation and deliberation at different levels, others recognize this as a threat to rapid economic development that will slow down decision-making.

This is evident in a post-2014 election India with the National Democratic Alliance (NDA) in power with a mandate of rapid economic growth, stealthily dismantling spaces for civil society to critically engage with policy decisions. One such move was to constitute a National Board for Wildlife (NBWL) with almost no civil society representation, which eases forest clearances for industrial projects. This has come in the wake of fast-tracking the process of obtaining clearances for a number of industrial and development projects: 240 projects in three months, from mining to roads. This makes it impossible to undertake proper environmental impact studies, public hearings and other mandated procedures (*Hindustan Times* 2014). More pertinent to the case studies in this book, the NDA government through the MoEF has also issued a letter diluting the provisions of the FRA, allowing corporations to get consent for diverting forestland for industrial projects from state-appointed district collectors rather than village councils who were empowered with the task and often denied requests. This has led to resistance by tribal activists and environmentalists who have urged the government not to "follow development at [the] expense of rights of tribals" (*Daily News and Analysis* 2014). One marked feature of the inclusion of social movements and extra-parliamentary advocacy is the engagement of stakeholders beyond the formulation stage, through processes of implementation. The divisive issues that have been identified in contentious deliberation play out in many different sites and strategies in the implementation phase. Follow-up research would be useful to see how advocacy and deliberation play out over time and to what extent ideas accepted into policy are mainstreamed into implementation policies. It would also be useful to study if these policies empower the marginalized to take on the state or industry in local sites of deliberation and what strategies are appropriated at that level.

What we see emerging in the wake of these two policies is the continued narrative conflict between the stances of ecological modernism of the state with civic environmentalism emphasizing equity and justice. These narratives are becoming increasingly polarized by the "pro-growth" agenda of the NDA government, while the role of industry has been markedly absent in the two cases presented, especially at the formulation stage. One will find that the dynamics of participation will be markedly different as industry actively engages in the implementation processes. This coupled with the aggressive "investor friendly" stance of the current government, its strategic crackdown on dissent and the removal of critical independent voices from institutions of environmental governance will nuance the dynamics of both advocacy and deliberation. One may see, for instance, the closing up of critical invited spaces but an expansion of invented spaces through extra-parliamentary advocacy and social movements. What is clear is that the fragile role of participation within a deliberative sphere can be buffered only by civil society activities in the public space scrutinizing and protecting the narrow deliberative spaces that keep critical channels of communication with the state open and receptive. This underlines the key message of this book that advocacy in extra-parliamentary spaces is a necessary aspect to inclusion in the more formal, parliamentary, deliberative spaces and that continued activism is necessary for outcomes in the formal deliberative process. It points to the idea that nascent spaces of participation, allowing access and voice, are constantly threatened with marginalization and that advocacy guards these spaces against cooptation.

While these narratives have been presented as coherent blocks in this book and have had an important role to play in creating self-aware identities, in the implementation phase of the two policies, these narratives are more fragmented with new definitions entering the fray. For example, a tribal and non-tribal, even if both are dependent on forest resources, have different interpretations of their rights on the forest. In implementation, identity politics and the ideas associated with what it means to be a tribal or non-tribal will strengthen or weaken certain claims on the land. More pertinently, the definitions created in this law, along with the cut-off dates for who can file a claim to the land, become contentious in the way the government views these differing claims. The problems of definition that persist with defining forests and biodiversity play out in several ways in the implementation of these laws. The lack of coherence in vocabulary about the resources in question leads to a conflict and contestation about the meanings imbued to them. In this these resources become particularly vulnerable to vested interests.

One of the defining features of the two policies was encoding consultation in the law. For example the FRA (along with the Wildlife Protection Amendment Act, 2006) makes provisions for consulting with people and obtaining their consent in the process of declaring tiger reserves and critical wildlife habitats (CWHs). This is an important recognition of local people's rights over their natural resources and has come into conflict with the fines-and-fences approach that was the norm. For instance, by the spring of 2003 almost 500 villages within India, occupied by a total of 300,000 people, had experienced forced relocation to protect the habitats of wildlife by exclusionary measures (Dowie 2009). This has highlighted the issue of marginalization both in the formulation process and in its implementation. As discussed, marginalization is a key frame in these policies and operates on several levels. A lack of voice keeps people marginalized from the policy process, and, on another level, people continue to be marginalized from the material resource in itself. These policies attempted to address this marginalization and, in opening up the space to different voices, created a more self-aware identity. The re-iteration of historical wrongs being righted through these policies allowed for a far more assertive demand for rights that continues into their roles in implementing these policies. In this the policies themselves continue to be living processes, used in different ways to counter different demands and claims on the land.

It seems most probable that the growing pressures on land in India will only escalate. Population growth, economic growth and diversion and encroachment of forestland for industry and agriculture are all growing problems. For example, twice as many diversions of forest for mining were granted from 1997 to 2007 than for the previous ten years (Nayak et al. 2008). In addition there is a trend of increasing use of land for export crops rather than growing food for local consumption (Patnaik 2007). Democratic solutions are required to negotiate solutions between the different claims on land, and this process of formulation shows the uneasy relationship of the state with democratic decision-making. The tensions between a "scientific" approach to forest and biodiversity protection is even now coming into conflict with more people-centered approaches. Ultimately the test lies in the way the state is able to negotiate different demands, both local and global and the role of people in building strong advocacy movements that build genuinely democratic institutions rather than going through the motions of a staged democracy.

Notes

1 Government of India, Ministry of Environment, Forests and Climate Change, office memorandum no. J-IIOI5/30/2004-IA.I1 (M), July 28, 2014. See www.moef.nic.in/sites/default/files/OM%20dated%2028.07.2014%20one%20time%20expansion.pdf.
2 See http://indianexpress.com/article/india/india-others/govt-set-to-consult-public-on-all-new-laws-amendments/.
3 See Guidelines of the United Nations Office of the High Commissioner for Human Rights, www.ohchr.org/EN/Pages/WelcomePage.aspx.
4 Minutes from the National Forum of Forest People and Forest Workers National Convention, June 10–12, 2009, Dehradun, India (personal communication).

References

Abelson, J., P.-G. Forest, J. Eyles, P. Smith, E. Martin, and F. P. Gauvin. "Deliberations about Deliberative Methods: Issues in the Design and Evaluation of Public Participation Processes." *Social Science and Medicine* 57 (2003): 239–51.

Acharya, A. "How Ideas Spread: Whose Norms Matter? Norm Localization and Institutional Change in Asian Regionalism." *International Organization* 58 (2004): 239–75.

Acharya, K. "Biodiversity Policy: Bold, New Step." *Deccan Herald* (Bangalore), November 29, 2002. http://listi.jpberlin.de/pipermail/info-mail/2002-December/000048.html (accessed April 6, 2012).

———. "Biodiversity-India: Lax Laws Worry Villagers, Activists." *IPS Inter Press Service*, May 7, 2007. http://ipsnews.net/print.asp?idnews=37634 (accessed April 6, 2012).

Acharya, K.P. "Twenty-Four Years of Community Forestry in Nepal." *International Forestry Review* 42 (2002): 149–56.

Ackerman, J. "Co-governance for Accountability: Beyond 'Exit' and 'Voice.'" *World Development* 32, no. 3 (2004): 447–63.

Adler, E. "Constructivism and International Relations." In *Handbook of International Relations*, edited by W. Carlsnaes, B. Simmons, and T. Risse, 95–118. London: Sage, 2000.

Agarwal, A. "An Indian Environmentalist's Credo." In *Social Ecology*, 346–86. New Delhi: Oxford University Press, 1994.

Agrawal, A. "Dismantling the Divide between Indigenous and Scientific Knowledge." *Development and Change* 26 (1995): 413–39.

Ali, A. "The Evolution of the Public Sphere in India." *Economic and Political Weekly*, June 30, 2001.

Anand, N. "Planning Networks: Processing India's National Biodiversity Strategy and Action Plan." *Conservation and Society* 4 (2006): 471–87.

Ananthakrishnan, G. "Forest Commission Rips Apart Tribal Bill." *Indian Express*, June 18, 2006. www.indianexpress.com/news/forest-commission-rips-apart-tribal-bill/6808/1 (accessed May 15, 2012).

Andersen, R. *Governing Agrobiodiversity: Plant Genetics and Developing Countries.* Aldershot: Ashgate, 2008.

Andrews, K., and B. Edwards. "Advocacy Organizations in the U.S. Political Process." *Annual Review of Sociology* 30 (2002): 479–506.

Angelsen, A., and S. Wunder. "Exploring the Forest – Poverty Link: Key Concepts, Issues and Research Implications." CIFOR Occasional Paper, no. 40 (2003).

Anuradha, R. V., B. Taneja, and A. Kothari. "Experiences with Biodiversity Policy-Making and Community Registers in India." *Participation in Access and Benefit Sharing Series*, 5–57. London: IIED, 2001.

Apte, T. *An Activist Approach to Biodiversity Planning: A Handbook of Participatory Tools Used to Prepare India's National Biodiversity Strategy and Action Plan*. London: IIED, 2005.

————. "A Peoples' Plan for Biodiversity Conservation: Creative Strategies That Work (and Some That Don't)." *Gatekeepers Series 130*. London: IIED, 2006.

Arendt, H. *On Violence*. New York: Harcourt, Brace & World, 1970.

Arnstein, S. "A Ladder of Citizen Participation." *Journal of the American Institute of Planners* 35, no. 4 (1969): 216–24.

Arts, B., M. Appelstrand, D. Kleinschmit, H. Pülzl, I. Visseren-Hamakers, R. Eba'a Atyi, T. Enters, K. McGinley, and Y. Yasmi. "Discourses, Actors and Instruments in International Forest Governance." In *Embracing Complexity: Meeting the Challenges of International Forest Governance*, edited by J. Rayner, A. Buck, and P. Katila, 57–73. A Global Assessment Report Prepared by the Global Forest Expert Panel on the International Forest Regime. Vienna: International Union of Forest Research Organizations, 2010.

Bäckstrand, K., and E. Lövbrand. "Planting Trees to Mitigate Climate Change: Contested Discourses of Ecological Modernization, Green Governmentality and Civic Environmentalism." *Global Environmental Politics* 6, no. 1 (2006): 50–75.

Badavan, L. "MoEF Trying to Divert Attention from its Inaction." *Frontline*, February 10, 2006.

Ban, C. "Economic Ideas in Translation: Theoretical and Methodological Reconsiderations." Presented at the Alliance for Governance Research and Analysis (AGORA), June 12, 2010.

Ban, R., S. Jha, and V. Rao. "Who Has Voice in a Deliberative Democracy? Evidence from Transcripts of Village Parliaments in South India." *Journal of Development Economics*, forthcoming. Stanford Graduate School of Business Paper no. 2103. http://ideas.repec.org/p/ecl/stabus/2103.html (accessed September 10, 2012).

Barabas, J. "How Deliberation Affects Policy Opinions." *American Political Science Review* 98, no. 4 (2004): 687–701.

Bardhan, P. *Land, Labor and Rural Poverty*. New York: Columbia University Press, 1984.

Barnett, M. N. "Culture, Strategy and Foreign Policy Change: Israel's Road to Oslo." *European Journal of International Relations* 5 no. 1 (1999): 5–36.

Bass S., B. Dalal-Clayton B., and J. Pretty. "Participation in Strategies for Sustainable Development." *IIED Environmental Planning Issues*, no. 7 (1995).

Batliwala, S. "Grassroots Movements as Transnational Actors – Implications for Transnational Civil Society." *Voluntas* 13, no. 4 (December 2002): 393–409.

————. "When Rights Go Wrong – Distorting the Rights Based Approach to Development." (2010). www.justassociates.org/sites/justassociates.org/files/whenrightsgowrong.pdf (accessed July 25, 2012).

Bavadam, L. "MoEF Trying to Divert Attention from Its Inaction." *Frontline*, January 28, 2006.

Baviskar, A. *In the Belly of the River: Tribal Conflict over Development in the Narmada Valley*. New Delhi: Oxford University Press, 1995.

————. "Tribal Politics and Discourses of Environmentalism." *Contributions to Indian Sociology* 31, no. 2 (1997): 195–223.

Baxi, U. *Inhuman Wrongs and Human Rights: Unconventional Essays*. New Delhi: Har Anand Publications, 1994.

Bedsworth, L. W., M. D. Lowenthal, et al. "Uncertainty and Regulation: The Rhetoric of Risk in the California Low-Level Radioactive Waste Debate." *Science, Technology & Human Values* 24 (2004): 406–27.

Benford, R. D., and D. A Snow. "Framing Processes and Social Movements: An Overview and Assessment." *Annual Review of Sociology* 26 (2000): 611–39.

Benford, R. D., and S. Hunt. "Dramaturgy and Social Movements: The Social Construction and Communication of Power." *Sociological Inquiry* 62 (1992): 36–55.

Benhabib, S. "Deliberative Rationality and Models of Democratic Legitimacy." *Constellations* 1, no. 1 (1994): 26–52.

Berger, P. L., and T. Luckmann. *The Social Construction of Reality*. Garden City: Doubleday, 1967.

Bhatnagar, B., and A. Williams. "Participatory Development and the World Bank: Potential Directions for Change." World Bank Discussion Paper 183. Washington, DC, 1992.

Bhushan, C. "Rich Lands, Poor People: The Socio-environmental Challenges of Mining in India." 6th Citizen's Report. New Delhi: Centre for Science and Environment, 2008.

Bhutani, S., and A. Kothari. "The Biodiversity Rights of Developing Nations: A Perspective from India." *Golden Gate University Law Review* 32, no. 4 (2002). http://digital commons.law.ggu.edu/ggulrev/vol32/iss4/6.

Bijoy, C. R. "Access and Benefit Sharing from the Indigenous Peoples' Perspective: The TBGRI-Kani Model." *Law, Environment and Development Journal* 3, no. 1 (2007). www.lead-journal.org/content/07001.pdf.

———. "The Adivasis of India – A History of Discrimination, Conflict, and Resistance." People's Union for Civil Liberties. www.pucl.org/Topics/Dalit-tribal/2003/adivasi.htm (accessed November 20, 2011).

Bindra, P. S. "UPA's Tribal Bill: Tiger's Death Warrant." *Pioneer*, April 2005. http://india-environmentportal.org.in/content/211289/upas-tribal-bill-tigers-death-warrant/ (accessed June 15, 2012)

Biodiversity Convention: Convention on Biological Diversity, Rio de Janeiro. International Legal Materials (June 5, 1992): 31–818.

Biological Diversity Act. Government of India, 2002.

Birkland, T. A. 1997. *After Disaster: Agenda Setting, Public Policy and Focusing Events*. Washington, DC: Georgetown University Press, 1997.

———. "The Historical and Structural Context of Public Policy Making." In *An Introduction to the Policy Process: Theories, Concepts, and Models of Public Policy Making*, 25–51. New York: M. E. Sharpe, 2005.

Blakeney, M. "Bioprospecting and the Protection of Traditional Medical Knowledge of Indigenous Peoples: An Australian Perspective." *EIPR* 6 (1997): 298–303.

Blakesley, C. L. "Extraterritorial Jurisdiction." In *II International Criminal Law*, 2nd ed., edited by M. C. Bassiouni, 33–107. Ardsley, NY: Transnational Publishers, 1999.

Bobbio, N. *The Future of Democracy*. Translated by R. Griffen. Minneapolis: University of Minnesota Press, 1987.

Bogdan, R., and S. K. Biklen. *Qualitative Research for Education: An Introduction to Theory and Methods*. Boston: Allyn & Bacon, 1992.

Bogdan, R., and S. J. Taylor. *Introduction to Qualitative Research Methods: A Phenomenological Approach to the Social Sciences*. New York: Wiley, 1975.

Bohman, J. "The Coming of Age of Deliberative Democracy." *Journal of Political Philosophy* 6, no. 4 (1998): 400–425.

Bonn Guidelines on Access to Genetic Resources and Fair and Equitable Sharing of the Benefits Arising Out of Their Utilization. 2002. Bonn. UN Doc. UNEP/CBD/COP/6/20. http://archive.defra.gov.uk/environment/biodiversity/geneticresources/documents/bonnguidelines.pdf.

Bose, I. "How Did the Indian Forest Rights Act, 2006, Emerge? Improving Institutions for Pro-poor Growth." Department for International Development. 2010. Discussions paper series 39. www.ippg.org.uk/papers/dp39.pdf (accessed July 19, 2012).

Boyatzis, R. *Transforming Qualitative Information: Thematic Analysis and Code Development.* Thousand Oaks, CA: Sage, 1998.

Boyle, M., J. Kay, and B. Pond. "Monitoring in Support of Policy: An Adaptive Ecosystem Approach." In *Encyclopedia of Global Environmental Change*, edited by T. Munn, 4:116–37. London: John Wiley and Son, Canada, 2001.

Brahm, E. "Sovereignty: Beyond Intractability." *Conflict Research Consortium: University of Colorado.* Edited by G. Burgess and H. Burgess. 2004. Available at www.beyondintractability.org/essay/sovereignty/.

Bratton, M. "NGOs in Africa: Can They Influence Public Policy?" *Development and Change* 27, No. 4 (1988) 87–118.

Bridgman, T., and D. Barry. "Regulation Is Evil: An Application of Narrative Policy Analysis to Regulatory Debate in New Zealand." *Policy Sciences* 35, no. 1 (2002): 141–61.

Brody, B. A. "Intellectual Property, State Sovereignty, and Biotechnology." *Kennedy Institute of Ethics Journal* 20, no. 1 (2010), 51–73.

Broome, N. P. "India's Culture of Conservation." *InfoChange*, December 2011. http://infochangeindia.org/environment/backgrounder/india-s-culture-of-conservation.html (accessed June 15, 2012).

Brulle, R. J. "The U.S. Environmental Movement." In *20 Lessons in Environmental Sociology*, edited by K. Gould and T. Lewis. New York: Oxford University Press, 2008.

Brulle, Robert. "Politics and the Environment." In Handbook of Politics, edited by Kevin Leicht and Craig Jenkins 385–406. New York: Springer, 2010.

Brysk, A. "From Above and Below: Social Movements, the International System and Human Rights in Argentina." *Comparative Political Studies* 26 (October 1993): 259–85.

Buch, M. N. "Killing the Inheritors of the Earth." *Indian Express*, May 27, 2005.

Canessa, A. "Who Is Indigenous? Self-Identification, Indigeneity, and Claims to Justice in Contemporary Bolivia." *Urban Anthropology* 36, no. 3 (2007): 14–48.

Chalmers, D., S. Martin, and K. Piester. "Associative Networks: New Structures of Representation for the Popular Sectors?" In *The New Politics of Inequality in Latin America*, edited by D. Chalmers et al., 543–82. Oxford: Oxford University Press, 1997.

Chambers, S. "Deliberative Democratic Theory." *Annual Review of Political Science* 6 (2003): 307–26.

Charney, J. "Universal International Law." *American Journal of International* Law 87 (October 1993): 543–50.

Chatterjee, P. "On Civil and Political Society in Post-colonial Democracies." In *Civil Society: History and Possibilities*, edited by Sudipta Kaviraj and Sunil Khilnani, 165–78. Cambridge: Cambridge University Press, 2001.

Checkel, J. T. *Ideas and International Political Change: Soviet/Russian Behavior and the End of the Cold War.* New Haven: Yale University Press, 1997.

———. "Norms, Institutions, and National Identity in Contemporary Europe." *International Studies Quarterly* 43 (1999): 84–114.

———. "The Europeanization of Citizenship." In *Transforming Europe: Europeanization and Domestic Change*, edited by M. Green Cowles, J. Caporaso, and T. Risse, 180–97. Ithaca: Cornell University Press, 2001.

Cheria, A., C. R. Bijoy, and C. Edwin. *A Search for Justice: A Citizens Report on the Adivasi Experience in South India.* Bangalore: St Pauls Publications, 1997.

Chetkovich, C., and F. Kunreuther. *From the Ground Up: Grassroots Organizations Making Social Change.* New York: ILR Press, 2006.

Chopra, D. "Policy Making in India: A Dynamic Process of Statecraft." *Pacific Affairs* 84, no. 1 (2011): 89–107.

Chowdhury A. "A Case Study of the National Forum of Forest People and Forest Workers." Unpublished, 2009.

Cicero. *Oratorium.* Loeb Classical Library. Cambridge, MA: Harvard University Press, 1942.

Clandinin, D. J., and F. M. Connelly. *Narrative Inquiry: Experience and Story in Qualitative Research.* San Francisco: Jossey-Bass, 2000.

Clark, J. *Democratizing Development: The Role of Voluntary Organizations.* West Hartford, CT: Kumarian Press, 1991.

———. "The State, Popular Participation, and the Voluntary Sector." *World Development* 23, no. 4 (1995): 593–601.

Clayton, A., ed. *NGOs, Civil Society and the State.* Oxford: Intrac, 1996.

Cohen, J., and A. Arato. *Civil Society and Political Theory.* Cambridge, MA: MIT Press, 1992.

Cohen, J., and J. Rogers. "Secondary Associations and Democratic Governance." *Politics and Society* 20, no. 4 (1992): 393–472.

Committee on Trade and Environment Council for Trade-Related Aspects of Intellectual Property Rights. "India and WTO" 2000, WT/CTE/W/156 IP/C/W/198 14 July 2000 http://www.wto.org/english/tratop_e/trips_e/t_agm3_e.htm#1. (Accessed May 29, 2012.)

Common Minimum Programme. Government of India, 2004. http://architexturez.net/system/files/pdf/cmp_0.pdf (accessed April 13, 2012).

Conover, P. "Identity, Emotion and Reason in the Same-Sex Marriage Debates." Paper Presented at the Annual Meetings of the American Political Science Association, Philadelphia, August 28–31, 2003.

Convention on Biological Diversity (CBD). 1992. Available at www.cbd.int.

———. COP 7 Decision VII/28: Protected Areas, 2004.

Conway, J. "Social Forums, Social Movements and Social Change: A Response to Peter Marcuse on the Subject of the World Social Forum." *International Journal of Urban and Regional Research* 29, no. 2 (2005): 425–28.

Cooke, B., and U. Kothari, eds. *Participation: The New Tyranny?* New York: Palgrave Macmillan, 2004.

Corkery J., A. Land, and J. Bossuyt. "The Process of Policy Formulation: Institutional Path or Institutional Maze?" European Centre for Development Policy Management, Policy Management reform, December 1995.

Cornwall, A. "Making Spaces, Changing Places: Situating Participation in Development." Brighton (England): Institute of Development Studies, 2002. vol.170. 35.

———. "Spaces for Transformation? Reflections on Issues of Power and Difference in Participation and Development." In *Participation: From Tyranny to Transformation,* edited by S. Hickey and G. Mohan, chap. 5. London: Zed Books 2004.

Cornwall, A., and V. S. Coelho, eds. *Spaces for Change? The Politics of Citizen Participation in New Democratic Arenas.* London: Zed Books, 2005.

Cortell, A., and J. Davis. "How Do International Institutions Matter: The Domestic Impact of International Rules and Norms." *International Studies Quarterly* 40 (1996): 451–78.

———. "Understanding the Domestic Impact of International Norms: A Research Agenda." *International Studies Review* 2, no. 1 (2000): 65–87.

Covey, J. G. "Accountability and Effectiveness of NGO Policy Alliances." IDR Reports, vol. 11, no. 8, 1994.

Creswell, J. W. *Qualitative Inquiry and Research Design: Choosing among Five Designs.* Thousand Oaks, CA: Sage, 1998.

———. *Research Design: Qualitative, Quantitative, and Mixed Methods Approaches.* Thousand Oaks, CA: Sage Publications, 2003.

Crosby, N. "Citizens Juries: One Solution for Difficult Environmental Questions." In *In Fairness and Competence in Citizen Participation,* edited by O. Renn, T. Webler, and P. Wiedemann, 157–74. Dordrecht, Netherlands: Kluwer Publishers, 1995.

Crouch, C. *Postdemocrazia.* Laterza: Roma-Bari.

CSD (Campaign for Survival and Dignity). *Endangered Symbiosis: Evictions and India's Forest Communities.* Report of the Jan Sunwai, July 19–20, 2003.

———. "Forest Rights Update: Forests Protests and Forest Wars." Call for All India Protests. New Delhi, 2005.

———. Tribals and Forest Dwellers Begin Forest Rights Dharna. Message posted on November 21, 2006, to http://groups.yahoo.com/group/tribal-India/message/1115.

———. Open Letter to Vanashakti, Open Letter to Wildlife Conservation Groups on Supreme Court Petition. Campaign for Survival and Dignity (n.d.).

Cullet, P., and J. Raja. "Intellectual Property Rights and Biodiversity Management: The Case of India." *Global Environmental Politics* 4, no. 1 (2004): 97–114.

Daily News and Analysis. "What Is a Forest? Environment Ministry Doesn't Know!" June 29, 2011. www.dnaindia.com/india/report_what-is-a-forest-environment-ministry-doesn-t-know_1560440 (accessed February 14, 2012).

———. "Activists Write to PM on the Dilution of the Forest Rights Act." November 4, 2014.

D'Arcy, S. "Deliberative Democracy, Direct Action, and Animal Advocacy." *Journal for Critical Animal Studies* 5, no. 2 (2007): 1–16.

Das, D. "Coordinated Programme on Biodiversity Soon." *Times of India,* April 15, 2011.

Das, K. "Combating Biopiracy, the Legal Way." *India Together,* May 6, 2005. www.india together.org/2005/may/env-biopiracy.htm (accessed May 29, 2012).

Dasgupta, B. *The Naxalite Movement.* New Delhi: Allied Publishers, 1974.

Dawkins, K. *Gene Wars: The Politics of Biotechnology.* New York: Seven Stories Press, 1997.

Deliberative Democracy Consortium (DDC). "Where Is Democracy Headed? Research and Practice on Public Deliberation." Kettering Foundation and Deliberative Democracy Consortium, 2008. http://americaspeaks.org/wpcontent/_data/n_0001/resources/live/Where_is_Democracy_Headed_report.pdf (accessed September 15, 2012).

Della Porta, D., and M. Diani. *Social Movements: An Introduction.* Oxford: Blackwell Publishers, 1999.

Denzin, N., and Y. Lincoln, eds. *Handbook of Qualitative Research.* London: Sage, 2000.

Deo, N., and D. McDuie-Ra. *The Politics of Collective Advocacy in India: Tools and Traps.* Sterling, VA: Kumarian Press, 2011.

de Tray, D. "Policy Impacts on the NGO Sector." Memo to Aubrey Williams in Response to Brown Paper, July 27. World Bank, Washington, DC, 1990.

Devraj, R. "Biodiversity Bill Passed." *India Together,* December 2002. http://indiatogether.org/environment/articles/biodiv02.htm (accessed June 5, 2010).

Devullu, P., et al. "Indigenous and Tribal Communities, Biodiversity Conservation and the Global Environment Facility in India – General Overview and a Case Study of People's Perspectives of the India Ecodevelopment Project." Samata, Hyderabad, 2004. www.forestpeoples.org.

Dewey, J. "The Quest for Certainty: A Study of the Relation of Knowledge and Action, Minton, Balch, and Company." New York, 1929. Reprinted in *John Dewey: The*

Later Works, 1925–1953; Volume 4; 1929, edited by J. Boydston, H. Furst Simon and Stephen Toulmin, 1–254. Carbondale and Edwardsville: Southern Illinois University Press, 1984.

———. "Liberalism and Social Action." In *John Dewey: The Later Works, 1925–1953; Volume 2; 1935–1937*, Edited by J. Boydston, H. Furst Simon and S. Toulmin 1–21. Carbondale and Edwardsville: Southern Illinois University Press, 1986.

Diamond, L. *Developing Democracy: Toward Consolidation*. Baltimore: Johns Hopkins University Press, 1999.

Divan, S., and A. Rosencranz. *Environmental Law and Policy in India: Cases*. New Delhi: Oxford University Press, 2001.

Dobson, C. *The Troublemaker's Teaparty: A Manual for Effective Citizen Action*. Gabriola Island, BC: New Society Publishers, 2003.

Dowie, M. *Conservation Refugees: The HundredYear Conflict between Global Conservation and Native Peoples*. Cambridge, MA: MIT Press.

Down to Earth. "Deep in the Woods." January 15, 2003.

Downes, D. R. "Global Trade, Local Economies, and the Biodiversity Convention." In *Biodiversity and the Law*, edited by W. J. Snape III and O. A. Houcj 202–216. Washington, DC: Island Press, 1996.

Drèze, J., and A. Sen. *India: Economic Development and Social Opportunity*. Delhi: Oxford University Press, 1995.

———, eds. *India: Development and Participation*. Delhi: Oxford University Press, 2002.

Drèze, J., M. Samson, and S. Singh. *The Dam and the Nation: Displacement and Resettlement in the Narmada Valley*. New Delhi: Oxford University Press, 1997.

Dryzek, J. S. *Discursive Democracy: Politics, Policy, and Political Science*. Cambridge, MA: Cambridge University Press, 1990.

———. *Deliberative Democracy and Beyond*. Oxford: Oxford University Press, 2000.

———. "Democratization as Deliberative Capacity Building." *Comparative Political Studies* 42, no. 11 (2009): 1379–1402.

———. *Foundations and Frontiers of Deliberative Governance*. Oxford: University Press, 2010.

Dryzek, J. S., and H. Stevenson. "Global Democracy and Earth System Governance." *Ecological Economics* 70, no. 11 (2011): 1865–74.

Dubey, S. "An Undemocratic Environment." *India Together*, October 19, 2006. www.india together.org/2006/oct/env-democenv.htm (accessed October 8, 2009).

Duraiappah, A. K. "Poverty and Environmental Degradation: A Review and Analysis of the Nexus." *World Development* 26, no. 12 (1998): 2169–79.

Dwivedi, O. P., and R. Khator. "India's Environmental Policy, Programs, and Politics." In *Environmental Policies in the Third World: A Comparative Analysis*, edited by O. P. Dwivedi and D. K. Vajpeyi, 47–71. Westport: Greenwood Press, 1995.

Dwivedi, R. "People's Movements in Environmental Politics: A Critical Analysis of the Narmada Bachao Andolan in India." Working Paper Series no. 242, Institute of Social Studies, 1997.

Eade, D., and E. Ligteringen, eds. *Debating Development: NGOs and the Future*. Oxford: Oxfam, 2001.

Economic and Political Weekly. "Forests and Tribals: Restoring Rights." January 6, 2007. http://www.epw.in/editorials/forests-and-tribals-restoring-rights.html (accessed September 23, 2015)

Eden, S. "Environmental Issues: Knowledge, Uncertainty and the Environment." *Progress in Human Geography* 22, no. 3 (1998): 425–32.

Edwards, M., and J. Gaventa, eds. *Global Citizen Action*. Boulder, CO: Lynne Rienner, 2001.

Eisenhardt, K. M., and M. E. Graebner. "Theory Building from Cases: Opportunities and Challenges." *Academy of Management Journal* 50, no. 1 (2007): 25–32.

Eliasoph, N. *Avoiding Politics: How Americans Produce Apathy in Everyday Life*. Cambridge: Cambridge University Press, 1998.

Elster, J. "The Market and the Forum: Three Varieties of Political Theory." In *Contemporary Political Philosophy: An Anthology*, edited by R. Goodin and P. Pettit, 128–42. Oxford: Blackwell, 1997.

———. "Deliberation and Constitution Making." In *Deliberative Democracy*, edited by J. Elster, 97–122. Cambridge: Cambridge University Press, 1998.

Elwin, V. *A New Deal for Tribal India*. New Delhi: Ministry of Home Affairs, GOI, 1939.

Escobar, A. "Whose Knowledge, Whose Nature? Biodiversity, Conservation, and the Political Ecology of Social Movements." *Journal of Political Ecology* 5 (1998): 53–80.

Esteves, A. M., S. Motta, and L. Cox. "Civil Society versus Social Movements." *Interface* 1, no. 2 (2009): 1–21.

Evans, P. B., D. Rueschemeyer, and Th. Skocpol, eds. *Bringing the State Back In*. Cambridge: Cambridge University Press, 1985.

Ezell, M. *Advocacy in the Human Services*. Belmont, CA: Wadsworth/Thomson Learning, 2001.

Fernandes, W., and G. Menon. "Tribal Women and Forest Economy, Deforestation, Exploitation and Status Change." Indian Social Institute, New Delhi, 1987.

Ferree, M. M., W. A. Gamson, J. Gerhards, and D. Rucht. *Shaping Abortion Discourse: Democracy and the Public Sphere in Germany and the United States*. Cambridge: Cambridge University Press, 2002.

Finnemore, M. *National Interests in International Society*. Ithaca, NY: Cornell University Press, 1996.

Finnemore, M., and K. Sikkink. "International Norm Dynamics and Political Change." *International Organization* 52 (1998): 887–917.

Fischer, F. *Citizens, Experts, and the Environment*. Durham, NC: Duke University Press, 2000.

———. *Reframing Public Policy: Discursive Politics and Deliberative Practices*. Oxford: Oxford University Press, 2003.

———. "Participatory Governance as Deliberative Empowerment: The Cultural Politics of Discursive Space." *American Review of Public Administration* 36, no. 1 (2006): 19–40. http://arp.sagepub.com/cgi/content/abstract/36/1/19.

Fischer, F., and J. Forester, eds. *The Argumentative Turn in Policy Analysis and Planning*. Durham, NC: Duke University Press, 1993.

Fisher, W. *Human Communication as Narration: Toward a Philosophy of Reason, Value, and Action*. Columbia: South Carolina University Press, 1987.

Fisk, M. *The State and Justice: An Essay in Political Theory*. Cambridge, MA: Cambridge University Press, 1989.

Fishkin, J. S. *Democracy and Deliberation: New Directions for Democratic Reform*. New Haven: Yale University Press, 1991.

———. *The Voice of the People: Public Opinion and Democracy*. New Haven: Yale University Press, 1996.

Fishkin, J. S., T. Gallagher, L. Luskin, J. McGrady, I. O'Flynn, and D. Russel. "A Deliberative Poll on Education: What Provision Do Informed Parents in Northern Ireland Want?" 2007. http://cdd.stanford.edu/polls/nireland/ (accessed December 11, 2009).

Flinter, M. "Biodiversity: Of Local Commons and Global Commodities." In *Privatizing Nature*, edited by M. Goldmann, 144–66. London: Pluto, 1997.

Flynn, J. "Communicative Power in Habermasian Theory of Democracy." *European Journal of Political Theory* 3, no. 4 (2004): 433–54.

Font, J., and I. Blanco. "Procedural Legitimacy and Political Trust: The Case of Citizen Juries in Spain." *European Journal of Political Research* 46 (2007): 557–89.

Fox, J. "Vertically Integrated Policy Monitoring: A Tool for Civil Society Policy Advocacy." *Nonprofit and Voluntary Sector Quarterly* 30, no. 3 (2001): 616–27.

Fraser, N. "Rethinking the Public Sphere: A Contribution to the Critique of Actually Existing Democracy." *Social Text* 25–26 (1990): 56–80.

———. "Rethinking the Public Sphere: A Contribution to the Critique of Actually Existing Democracy." In *Justice Interruptus: Critical Reflections on the "Postsocialist Condition"*, 56–80. Routledge: London, 1997.

Freeman, S. "Deliberative Democracy: A Sympathetic Comment." *Philosophy and Public Affairs* 29, no. 4 (2000): 371–418.

Friedman, E. J., and K. Hochstetler. "Assessing the Third Transition in Latin American Democratization: Representational Regimes and Civil Society in Argentina and Brazil." *Comparative Politics* 35, no. 1 (2000): 21–42.

Fung, A. "Accountable Autonomy: Toward Empowered Deliberation in Chicago Schools and Policing." *Politics and Society* 29, no. 1 (2001): 73–104.

———. "Recipes for Public Spheres: Eight Institutional Design Choices and Their Consequences." *Journal of Political Philosophy* 11, no. 3 (2003): 338–67.

———. "Deliberation before the Revolution – Toward an Ethics of Deliberative Democracy in an Unjust World." *Political Theory* 33 (2005): 397–419.

Fung, A., and E. O. Wright, eds. *Deepening Democracy: Institutional Innovations in Empowered Participatory Governance*. London: Verso, 2003.

Gabardi, W. "Contemporary Models of Democracy." *Polity* 33, no. 4 (2001): 547–68.

Gadgil, M., and R. Guha. *Ecology and Equity: The Use and Abuse of Nature in Contemporary India*. Routledge: London, 1996.

Galanter, M. *Competing Equalities: Law and the Backward Classes in India*. Berkeley: University of California Press, 1984.

Ganapathy, N. "Ministry Strikes at Tribal Land Rights Bill." *Indian Express*, April 16, 2005.

Ganz, M. "What Is Organizing?" *Social Policy* 33, no. 1 (Fall 2002), 16–17.

Garg, A. "Orange Areas, Examining the Origin and Status, Advocacy Perspective." Working Paper Series no. 21. Pune: National Centre for Advocacy Studies, 2005.

Gastil, J. *Democracy in Small Groups: Participation, Decision Making, and Communication*. Philadelphia: New Society, 1993.

———. *By Popular Demand: Revitalizing Representative Democracy through Deliberative Elections*. Berkeley: University of California, 2000.

Gastil, J., and J. P. Dillard. "Increasing Political Sophistication through Public Deliberation." *Political Communication* 16 (1999): 3–23.

Gaventa, J. "Towards Participatory Governance: Assessing the Transformative Possibilities." Paper presented at conference Participation: From Tyranny to Transformation, Manchester, 2003.

Ghosh, J. "Saving Forests and People." *Frontline*, June 2005. www.frontline.in/static/html/fl2213/stories/20050701004711800.htm (accessed September 10, 2010).

Ghosh, S. "India: The Forest Rights Act, a Weapon for Struggle." *WRI's Bulletin*, February 2006.

Giddens, A. *Modernity and Self-Identity: Self and Society in Late Modern Age*. Stanford: Stanford University Press, 1991.

Glowka, L., T. Plan, and P.-T. Stoll. "Best Practices for Access to Genetic Resources." Background paper for the Correspondent Workshop on "Best Practices for Access to Genetic Resources." January 14–17, 1997, Cordoba.

Gopalakrishnan, S. "A Tale of Two Bills." Unpublished manuscript, April 2006.

———. "Rights Legislations and the Indian State: Understanding the Nature and Meaning of the Forest Rights Act." Briefing prepared for mass organizations. Distributed by SRUTI, 2010.

Government of India. "Report of the Commission for Scheduled Castes and Scheduled Tribes." Commission for Scheduled Castes and Scheduled Tribes, Controller of Publication. New Delhi, 1992.

———. Census of India: Census Data 2001; India at a Glance. Office of the Registrar General and Census Commissioner, India, 2001.

———. The Scheduled Tribes and Other Traditional Forest Dwellers (Recognition of Forest Rights) Act, 2006, no. 2 of 2007. The Gazette of India; Government of India, New Delhi.

———. "Development Challenges in Extremist Affected Areas." Report of an Expert Group to Planning Commission, 2008. http://planningcommission.gov.in/reports/publi cations/rep_dce.pdf (accessed September 18, 2010).

Green, M., and B. Houlihan. "Advocacy Coalitions and Elite Sport Policy Change in Canada and the United Kingdom." *International Review for the Sociology of Sport* 39, no. 4 (2004): 387–403.

Greene, S. *Customizing Indigeneity: Paths to a Visionary Politics in Peru.* Stanford: Stanford University Press, 2009.

Grindle, M. S. *Politics and Policy Implementation in the Third World.* Princeton, N.J.: Princeton University Press, 1980.

Grugel, J., and P. Enrique. "Grounding Global Norms in Domestic Politics: Advocacy Coalitions and the Convention on the Rights of the Child in Argentina." *Journal of Latin American Studies* 42 (2010): 29–57.

Guha, R. "Forestry in British and Post-British India: A Historical Analysis." *Economic and Political Weekly*, October 29–November 5–12, 1983.

———. "Radical American Environmentalism and Wilderness Preservation: A Third World Critique." *Environmental Ethics* 11 (1989): 71–83.

Gupta, A. "Blurred Boundaries: The Discourse of Corruption, the Culture of Politics, and the Imagined State." *American Ethnologist* 22, no. 2 (1995), 375–402.

Gutmann, A., and D. Thompson. *Democracy and Disagreement.* Cambridge, MA: Belknap Press, 1996.

———. *Why Deliberative Democracy?* Princeton: Princeton University Press, 2004.

Guzzini, S. "A Reconstruction of Constructivism in International Relations." *European Journal of International Relations* 6, no. 2 (2000): 142–82.

Haas, P. *Saving the Mediterranean: The Politics of International Environmental Cooperation.* New York: Columbia University Press, 1990.

Habermas, J. "New Social Movements." *Telos* 49 (1981): 33–37.

———. *The Theory of Communicative Action, Vol. 1.* Boston: Beacon Press, 1984.

———. *The Theory of Communicative Action, Vol. 2: Life World and System; A Critique of Functionalist Reason.* Boston: Beacon Press, 1987.

———. "Three Normative Models of Democracy." *Constellations* 1, no. 1 (1994): 1–10.

———. *Between Facts and Norms.* Cambridge: Polity, 1996.

———. "Concluding Comments on Empirical Approaches to Deliberative Politics." *Acta Politica* 40 (2005): 384–92.

Hagopian, F. "Democracy and Political Representation in Latin America in the 1990s: Pause, Reorganization, or Decline?" In *Fault Lines of Democracy in Post-Transitional Latin America*, edited by F. Agüero and J. Stark, 99–143. Boulder, CO: North-South Center Press/University of Miami, 1998.

Hajer, M. A. *The Politics of Environmental Discourse: Ecological Modernization and the Policy Process*. Oxford: Oxford University Press, 1995.

Hajer, M. A., and H. Wagenaar. "Introduction." In *Deliberative Policy Analysis: Understanding Governance in the Network Society*, edited by M. Hajer and H. Wagenaar, 1–32. Cambridge: Cambridge University Press, 2003.

Hammersley, M., and R. Gomm. "Introduction." In *Case Study Method: Key Issues, Key Texts*, edited by R. Gomm, M. Hammersley, and P. Foster, 1–16. London: Sage, 2000.

Hancock, T., R. Labonte, and R. Edwards. "Indicators That Count! – Measuring Population Health at the Community Level." Toronto: Centre for Health Promotion at University of Toronto and ParticipACTION, 2000.

Hardt, M., and A. Negri. *Empire*. Cambridge, MA: Harvard University Press, 2000.

Harvie, B. (2002). *Regulation of Advocacy in the Voluntary Sector: Current Challenges and Some Responses*. Ottawa: Voluntary Sector Initiative, Government of Canada, 2002.

Healey, P. "The Communicative Work of Development Plans." *Environment and Planning B: Planning and Design* 20 (1993): 83–104.

Hechter, M., and O. Karl-Dieter, eds. *Social Norms*. New York: Russell Sage Foundation, 2001.

Hegel, G. W. F. *Philosophy of Right*. Translated by T. M. Knox. Oxford: Oxford University Press, 1967 (1821).

Heller, P. "Moving the State: The Politics of Democratic Decentralization in Kerala, South Africa and Porto Alegre." *Politics and Society* 29, no. 1 (2001): 131–63.

Hendriks, C. M. "The Ambiguous Role of Civil Society in Deliberative Democracy." Paper presented to the Jubilee conference of the Australasian Political Studies Association, Australian National University, Canberra, 2002.

———. "Integrated Deliberation: Reconciling Civil Society's Dual Role in Deliberative Democracy." *Political Studies* 54 (2006a): 486–508.

———. "When the Forum Meets Interest Politics: Strategic Uses of Public Deliberation." *Politics and Society* 34, no. 4 (2006b): 571–602.

Hickey, S., and G. Mohan, eds. *Participation: From Tyranny to Transformation? Exploring New Approaches to Participation in Development*. London: Zed Books, 2004.

Hindu. "Accept JPC Report on ST Forest Rights Bill, Government urged." June 26, 2006a.

———. "JPC Report on Forest Rights Hailed." July 23, 2006b.

———. "My Intention Is to Ensure 8 Per cent Economic Growth: Chidambaram." September 11, 2006c.

———. "Tribal Rights Bill Passed in Lok Sabha." December 16, 2006d.

———. "Laws by the People, for the People." February 24, 2014.

Hindustan Times. "Prakash Javadekar Clears 240 Projects in 3 Months." September 11, 2014. www.hindustantimes.com/india-news/on-fast-track-environment-minister-prakash-javadekar-clears-240-projects-in-3-months/article1-1262676.aspx.

Holston, J., and A. Appadurai. "Cities and Citizenship." In *Cities and Citizenship*, edited by J. Holston, 1–18. Durham, NC: Duke University Press, 1999.

Hummel, R. "Critique of Public Space." *Administration & Society* 34, no. 1 (2002): 102–7.

Innes, J. E., and D. E. Booher. "Collaborative Policy Making: Governance through Dialogue." In *Deliberative Policy Analysis: Governance in the Network Society*, edited by M. W. Hajer and H. Wagenaar, 33–59. Cambridge: Cambridge University Press, 2003.

James, W. *Varieties of Religious Experience*. Edited by Frederick H. Burkhardt, Fredson Bowers, and Ignas K. Skrupskelis. Cambridge, MA: Harvard University Press, 1902.

Johansson-Stenman, O. "The Importance of Ethics in Environmental Economics with a Focus on Existence Values." *Environmental & Resource Economics* 113–14 (1998): 429–42.

Kabeer, N. " 'Growing' Citizenship from the Grassroots: Nijera Kori and Social Mobilization in Bangladesh." In *Inclusive Citizenship: Meanings and Expressions*, edited by N. Kabeer. London: Zed Books, 2005.

Kalpavriksh. "The Biological Diversity Act 2002 and Rules 2004: Concerns and Issues." http://wgbis.ces.iisc.ernet.in/biodiversity/sahyadri_enews/newsletter/issue27/pdfs/Biodiversity%20Act%20and%20Rules,%20basic%20note.pdf (accessed November 20, 2008).

Kameda, T. "Procedural Influence in Small Group Decision Making: Deliberation Style and Assigned Decision Rule." *Journal of Personality and Social Psychology* 61 (1991): 245–56.

Kant, S., and R. Cooke. "Jabalpur District, Madhya Pradesh, India: Minimizing Conflict in Joint Forest Management." International Development Research Centre (IDRC), Ottawa, 1999.

Kaplan, T. "Narrative Structure of Policy Analysis." *Journal of Policy Analysis and Management* 5, no. 4 (1986): 761–88.

Kathuria, V. "Informal Regulation of Pollution in a Developing Country: Evidence from India." *Ecological Economics* 63 (2007): 403–17.

Katzsenstein, M., S. Kothari and U. Mehta. 'Social movement politics in India: institutions, interests and identities' in in Kohli (ed.), The Success of India's Democracy (2001). Pp: 242–269.

Kaviraj, S., and S. Khilnani, eds. *Civil Society: History and Possibilities*. Cambridge: Cambridge University Press, 2001.

Keck, M.E., and K. Sikkink. *Activists beyond Borders: Advocacy Networks in International Politics*. Ithaca: Cornell University Press, 1998a.

———. "A Survey of Human-Rights Law." *Economist*, December 5, 1998b.

Khare, A. et al. *Joint Forest Management: Policy Practice and Prospects*. London: IIED, 2000.

Khator, R. *Environment, Politics and Development in India*. Lanham, MD: University Press of America, 1991.

Kitschelt, H. "Political Opportunity Structures and Political Protest: Anti-nuclear Movements in Four Democracies." *British Journal of Political Science* 16 (1986): 57–85.

Knight, J., and J. Johnson. "What Sort of Equality Does Deliberative Democracy Require?" In *Deliberative Democracy: Essays on Reason and Politics*, edited by J. Bohman and W. Rehg, 279–319. Cambridge, MA: MIT Press, 1997.

Kohli, A. *State Power and Social Forces: Domination and Transformation in the Third World*. New York: Cambridge University Press, 1994.

Kohli, K. "Biodiversity: Read the Fine Print." *India Together*, June 5, 2007. www.india together.org/2007/jun/env-biofinepr.htm (accessed March 11, 2012).

———. "NBSAP to NBAP: The Downward Spiral." *India Together*, March 30, 2009. www.indiatogether.org/2009/mar/env-nbap.htm (accessed May 10, 2010).

Kooiman, J. "Social-Political Governance: Overview, Reflections and Design." *Public Management* 1, no. 1 (March 1999): 67–92.

———. "Societal Governance: Levels, Models and Orders of Social-Political Interaction." In *Debating Governance: Authority, Steering and Democracy*, edited by J. Pierre 138–164. London: Sage, 2000.

Kothari, A. "Intellectual Property Rights and Biodiversity: Are India's Proposed Biodiversity Act and Plant Varieties Act Compatible?" Paper presented at Workshop on Biodiversity Conservation and Intellectual Property Regime, RIS/Kalpavriksh/IUCN, New Delhi, January 29–31, 1999.

————. "An Exercise in Conservation." *Frontline*, February 2, 2001. www.frontline.in/static/html/fl1802/18020670.htm (accessed June 5, 2011).

————. "Greening or Green Wash." In *Shades of Green: A Symposium on the Changing Contours of Indian Environmentalism*. No. 516, August. New Delhi: Seminar Publications, 2002.

————. "Draft National Environment Policy 2004: A Critique." *Economic and Political Weekly* 39, no. 43 (October 23, 2004): 4723–27.

————. India's Most Comprehensive Biodiversity Compilation. Message posted on September 4, 2005, to http://dir.groups.yahoo.com/group/ecwatch/message/42.

————. "Rights and Promises." *Frontline*, July 5, 2006.

Kothari, R. "Non-party Political Process." *Economic and Political Weekly* 19, no. 5 (1984): 216–24.

————. *State against Democracy: In Search of Humane Governance*. New York: New Horizons Press, 1989.

Kramer, R. M. "Nonprofit Organizations in the 21st Century: Will Sector Matter?" Aspen Institute, Nonprofit Sector Research Fund, Working Paper Series. Washington, DC, 1998.

Kriesi, H. "Political Context and Opportunity." In *The Blackwell Companion to Social Movements*, edited by D. A. Snow, S. A. Soule, and H. Kriesi, 67–90. Malden, MA: Blackwell Publishing, 2004.

Krishnan, R. "Forest Rights Act 2006 – Misplaced Euphoria." *Libération*, January 2007. www.cpiml.org/liberation/year_2007/January/forest_right_act.html (accessed December 6, 2009).

Kulkarni, S. "Forest Legislation and Tribals: Comments on Forest Policy Resolution." *Economic and Political Weekly*, December 12, 1987.

Kumar, A. "Violence and Political Culture: Politics of the Ultra-left in Bihar." *Economic and Political Weekly* 38, no. 47 (2003): 4977–83.

Kumar, V. A. "Policy Processes and Policy Advocacy." GAPS Series, Working Paper 7, 2006.

Kymlicka, W. *Contemporary Political Philosophy*. New York: Oxford University Press, 2002.

Larmore, C. "Pluralism and Reasonable Disagreement." Social Philosophy and Policy II (1): 61–79.

Leach, M., G. Bloom, A. Ely, P. Nightingale, I. Scoones, E. Shah, and A. Smith. "Understanding Governance: Pathways to Sustainability." STEPS Working Paper 2. Brighton: STEPS Centre, 2007.

Lefebvre, H. *The Production of Space*. London: Verso, 1991.

Lenin, J. "Conservation Hypocrisy: Again, They Come for the Forest Dwellers." *First Post*, October 30, 2011. www.firstpost.com/india/conservation-hypocrisy-again-they-come-for-the-forest-dwellers-119084.html (accessed June 6, 2012).

Lertzman, K., J. Rayner, and J. Wilson. "Learning and Change in the British Columbia Forest Sector: A Consideration of Sabatier's Advocacy Coalition Framework." *Canadian Journal of Political Science* 29 (1996): 11–134.

Lewis, J. P. *India's Political Economy: Governance and Reform*. Oxford: Oxford University Press, 1995.

Lewis, M. *Inventing Global Ecology: Tracking the Biodiversity Ideal in India, 1947–1997*. Athens, OH: Ohio University Press, 2004.

Lichterman, P. "Social Capital or Group Style? Rescuing Tocqueville's Insights on Civic Engagement." *Theory and Society* 35 (2006): 529–63.

Lincoln, Y. S., and E. Guba. *Naturalistic Enquiry.* Beverly Hills: Sage, 1985.

Linders, S. H., and Peters, B. G. "Instruments of Government: Perceptions and Contexts." *Journal of Public Policy* 9, no. 1 (1989): 35–58.

Madhusudan, M. D. "Of Rights and Wrongs: Wildlife Conservation and the Tribal Bill." *Economic and Political Weekly* 40, no. 47 (November 19, 2005): 4893–95.

Mahapatra, R., K. S. Shrivastava, S. Narayanan, and A. Pallavi. "How Government Is Subverting Forest Right Act." *Down to Earth*, 2010. New Delhi. www.downtoearth.org.in/coverage/how-government-is-subverting-forest-rights-act-2187 (accessed November 15, 2010).

Manin, B. "On Legitimacy and Political Deliberation." *Political Theory* 15, no. 3 (1987): 338–68.

———. *The Principles of Representative Government.* Cambridge: Cambridge University Press, 1997.

Mansbridge, J. J. *Beyond Adversary Democracy*. Chicago: University of Chicago Press, 1983.

———. "Everyday Talk in the Deliberative System." In *Deliberative Politics: Essays on Democracy and Disagreement*, edited by S. Macedo, 211–39. New York: Oxford University Press, 1999.

March, J. G., and J. P. Olsen. *Democratic Governance*. New York: Free Press, 1995.

Marcuse, P. "Are Social Forums the Future of Social Movements?" *International Journal of Urban and Regional Research* 29, no. 2 (2005): 417–24.

Marks, G., and D. McAdam. "On the Relationship of Political Opportunities to the Form of Collective Action: The Case of the European Union." In *Social Movements in a Socializing World*, edited by D. Della Porta, H. Kriesi, and D. Rucht, 97–111. London: McMillan Press, 1999.

Marshall, C., and G. B. Rossman. *Designing a Qualitative Research.* London: Sage Publications, 2008.

Martínez Novo, C. *Who Defines Indigenous: Identities, Development, Intellectuals and the State in Northern Mexico.* New Brunswick, NJ: Rutgers University Press, 2006.

Matthews, E., and Mock. G. "More Democracy, Better Environment." *World Resources* 4, (2002), 32–33.

Mathur, N., and K. Mathur. "Policy Analysis in India: Research Bases and Discursive Practices." In *Handbook of Public Policy Analysis Theory, Politics and Methods*, edited by F. Fischer et al., 603–16. Boca Raton, FL: CRC Press, 2007.

Mathur, V. B., and A. Rajvanshi. "Integrating Biodiversity into Environmental Impact Assessment: A National Case Study from India." Prepared under the Biodiversity Planning Support Programme of UNDP, UNEP and GEF, 2001. www.cbd.int/doc/nbsap/EIA/India.pdf (accessed December 6, 2011).

Mawdsley, E., D. Mehra, and K. Beazley. "Nature Lovers, Picnickers and Bourgeois Environmentalism." *Economic and Political Weekly* 44, no. 11 (March 14–20, 2009): 49–60.

McAdam, D. *Political Process and the Development of Black Insurgency.* Chicago: University of Chicago Press, 1982.

McAdam, D., J. D. McCarthy, and M. N. Zald. "Social Movements." In *Handbook of Sociology*, edited by N. J. Smelser, 695–737. Beverly Hills: Sage Publications, 1988.

———. *Comparative Perspectives on Social Movements: Political Opportunities, Mobilizing Structures, and Cultural Framings*. New York: Cambridge University Press, 1996.

McAdam, D., S. Tarrow, and C. Tilly. *Dynamics of Contention*. New York: Cambridge University Press, 2001.

McAfee, K. "Selling Nature to Save It? Biodiversity and Green Developmentalism." *Environment and Planning D* 17, no. 2 (1999): 133–54.

McBeth, M. K., E. A. Shanahan, R. J. Arnell, and P. L. Hathaway. "The Intersection of Narrative Policy Analysis and Policy Change Theory." *Policy Studies Journal* 35, no. 1 (2007): 87–108.

McBeth, M. K., E. A. Shanahan, et al. "Buffalo Tales: Interest Group Policy Stories in Greater Yellowstone." *Policy Sciences* 43, no. 4 (2010): 391–409.

McCarthy, J. D., and M. N. Zald. "Resource Mobilization and Social Movements: A Partial Theory." *American Journal of Sociology* 82 (1977): 1212–41.

McGee, R. "Unpacking Policy: Actors, Knowledge and Spaces." In *Unpacking Policy: Actors, Knowledge and Spaces in Poverty Reduction*, edited by K. Brock, R. McGee, and J. Gaventa, 1–26. Kampala: Fountain Press, 2004.

Mehta, A. K., and A. Shah. "Chronic Poverty in India: Incidence, Causes and Policies." *World Development* 31, no. 3 (2003): 491–511.

Menon, A. "Situating Law: The Political Economy of Environment and Development in India." In *Law, Land Use and the Environment: Afro-Indian Dialogues*, edited by C. Eberhard. Pondicherry, 363–386. Pondicherry: IFP, 2008.

Menon, S. "The Tragedy of Being Tribal in India." *Business Standard*, June 26, 2009. www.business-standard.com/article/opinion/sadanand-menon-the-tragedy-of-being-tribal-in-india-109062600012_1.html (accessed December 20, 2011).

Merriam, S. B. *Qualitative Research and Case Study Applications in Education*. San Francisco: Jossey-Bass, 1998.

Michael, M. "Lay Discourses of Science: Science-in-General, Science-in-Particular and Self." *Science Technology and Human Values* 17, no. 3 (1992): 313–33.

Michaelidou, M., and D. J. Decker. "European Union Policy and Local Perspectives: Nature Conservation and Rural Communities in Cyprus." *Cyprus Review* 15, no. 2 (2003): 121–45.

Miles, M. B., and A. M. Huberman. *Qualitative Data Analysis*. 2nd ed. Newbury Park, CA: Sage, 1994.

Miller, M. (1998). "Sovereignty Reconfigured: Environmental Regimes and Third World States." In *The Greening of Sovereignty in World Politics*, edited by K. Liftin, 172–92. Cambridge, MA: MIT Press, 1998.

Ministry of Food and Agriculture. National Forest Policy "Ministry of Food and Agriculture Resolution." May, 12 1952. http://forest.ap.nic.in/Forest%20Policy-1952.htm (accessed October 10, 2012).

Ministry of Home Affairs. "Co-ordination Centre Meeting on Naxalism Held." Press Information Bureau, Government of India. http://pib.nic.in/newsite/erelease. aspx?relid=27192 (accessed April 16, 2012).

Mitchell, T. "The Limits of the State: Beyond Statist Approaches and Their Critics." *American Political Science Review* 85 (March 1991): 77–96.

MoEF (Ministry of Environment and Forests, New Delhi, India). Forest (Conservation) Act, 1980, Forest (Conservation) Rules. Government of India, New Delhi, 1980.

———. National Forest Policy 1988. Government of India, New Delhi, 1988.

———. "Policy Statement on Abatement of Pollution." 1992. http://envfor.nic.in/down loads/about-the-ministry/introduction-csps.pdf.

———. The Environmental Impact Assessment Notification. Government of India, 1994 (amended on April 10, 1997).

———. *Implementation of Article 6 of the Convention on Biological Diversity in India – National Report.* New Delhi, 1998.

———. "From the Local to the Global: Developing the National Biodiversity Strategy and Action Plan." In *Towards Sustainability: Learning from the Past, Innovating for the*

Future. New Delhi, 2002. www.moef.nic.in/divisions/ic/wssd/doc3/chapter1/css/Chapter1.htm (accessed June 20, 2011).

———. Publication of the Biological Diversity Act, 2002 No. 18 of 2003. Ministry of Law and Justice (Legislative Department), Government of India, New Delhi, 2003.

———. *Securing India's Future: The National Biodiversity Strategy and Action Plan.* Draft manuscript. Government of India, 2004.

———. *Joining the Dots.* Report of the Tiger Task Force. New Delhi, 2005.

———. *Asia Pacific Forestry Sector Outlook Study I-Country Report.* New Delhi, 2009a.

———. *India's Fourth National Report to the Convention on Biological Diversity.* New Delhi, 2009b.

———. *Report to the People on Environment and Forest 2009–2010.* Annual report. Government of India, New Delhi, 2009c.

———. "NBSAP: India's experiences." Regional workshop for South, East and Southeast Asia on updating NBSAPs X'ian, China, May 9–16, 2011. www.cbd.int/doc/nbsap/nbsapcbw-seasi-02/NBSAP-Xian-India.pdf (accessed June 12, 2012).

Mohanty, B. "A Field View." *Seminar* 552 (August 2005): 30–34.

Mohanty, M., and A. K. Singh. *Voluntarism and Government: Policy, Programme and Assistance.* New Delhi: Voluntary Action Network India (VANI), 2001.

Montpetit, E., C. Rothmayr, and F. Varone. "Institutional Vulnerability to Social Constructions: Federalism, Target Populations, and Policy Designs for Assisted Reproductive Technology in Six Democracies." *Comparative Political Studies* 38 (2005): 119–42.

Montpetit, E., F. Scala, and I. Fortier. "The Paradox of Deliberative Democracy: The National Action Committee on the Status of Women and Canada's Policy on Reproductive Technology." *Policy Sciences* 37, no. 2 (2004): 137–57.

Mooij, J., and V. de Vos. "Policy Processes: An Annotated Bibliography on Policy Processes, with Particular Emphasis on India." Working Paper 221. London: Overseas Development Institute, 2003.

Mouffe, C. "Democracy, Power, and the Political." In *Democracy and Difference*, edited by S. Benhabib 245–255. Princeton: Princeton University Press, 1996.

———. "Deliberative Democracy or Agonistic Pluralism." *Political Sciences* 72 (2000a): 1–17.

———. *The Democratic Paradox.* London: Verso, 2000b.

———. *On the Political.* London: Routledge, 2005.

Mueller, C. M. "Conflict Networks and the Origins of Women's Liberation." In *New Social Movements*, edited by E. Lara, H. Johnston, and J. R. Gusfield, 234–63. Philadelphia: Temple University Press, 1994.

Mukul, A. "GoM Feels JPC Draft on Forest Bill May Not Help Tribals." *Times of India*, September 5, 2006.

Mutz, D. *Hearing the Other Side: Deliberative versus Participatory Democracy.* New York: Cambridge University Press, 2006.

Nash, K. *Contemporary Political Sociology: Globalization, Politics and Power.* Oxford: Blackwell, 2000.

National Forum of Forest People and Forest Workers (NFFPFW). "Forest Right Act: A Weapon of Struggle." Unpublished, personal correspondence, 2008.

National Forum of Forest People and Forest Workers National Convention (NFFPFW Convention). Conference circular, Dehradun, India, June 10–12, 2009.

Nayak, S. K., et al. "Economic Growth and Deforestation." In *Nought without Cause*, edited by M. Wani 110–127. Pune: Kalpavriksh, 2008.

NBSAP. Proceedings of the Inaugural National Workshop. New Delhi, June 23–24, 2000.

Neuman, W. L. *Social Research Methods: Qualitative and Quantitative Approaches*. 6th ed. Singapore: Pearson Education Incorporated, 2006.

Offe, C. "Reflections on the Institutional Self-Transformation of Movement Politics." In *Challenging the Political Order*, edited by R. Dalton and M. Kuechler, 232–50. Cambridge: Polity, 1990.

OHCHR (Office of the High Commissioner for Human Rights). "Human Rights and Poverty Reduction: A Conceptual Framework." 2004. www.ohchr.org/Documents/Publications/PovertyReductionen.pdf.

Ojha, H., and N. Timsina. "From Grassroots to Policy Deliberation – The Case of Federation of Forest User Groups in Nepal." In *Knowledge Systems and Natural Resources: Management, Policy and Institutions in Nepal*, edited by H. Ojha, N. Timsina, R. Chhetri, and K. Paudel 60–83. New Delhi: Cambridge University Press India and International Development Research Center, 2008.

Olson, M., Jr. *The Logic of Collective Action*. Cambridge, MA: Harvard University Press, 1965.

Osella, F., and C. Osella. *Social Mobility in Kerala: Modernity and Identity in Conflict*. London: Pluto Press, 2000.

Osman, F. A. "Public Policy Making: Theories and Their Implications in Developing Countries." *Asian Affairs* 24, no. 3 (2002), 37–53.

Osmani, S. R. "Participatory Governance: An Overview of the Issues Evidence." In *Participatory Governance and the Millennium Development Goals*, 1–48. New York: United Nations, 2007.

Ospina, S. M., and J. Dodge. "It's About Time: Catching Method Up to Meaning – The Usefulness of Narrative Inquiry in Public Administration Research." *Public Administration Review* 65, no. 2 (2005): 143–57.

Page, B. *Who Deliberates? Mass Media in Modern Democracy*. Chicago: University Chicago Press, 1996.

Pal, C. *Environmental Pollution and Development*. New Delhi: Mittal, 1999.

Panitch, L. "A Different Kind of State? In a Different Kind of State?" In *Popular Power and Democratic Administration*, edited by G. Albo, D. Langille, and L. Panitch, 2–17. Toronto: Oxford University Press, 1993.

Pant, M. "Social Mobilization and the State in India." India Country Paper written for IBSA comparative Synthesis project, 2010.

Panwar, T. S. "BJP, Congress in Himachal Follow Feudal Policies." *Him Vani*, April 14, 2009. www.himvani.com/news/2009/04/14/bjp-congress-in-himachal-follow-feudal-policies.

Papadopoulos, Y., and P. Warin. "Are Innovative, Participatory and Deliberative Procedures in Policy Making Democratic and Effective?" *European Journal of Political Research* 46 (2007): 445–72.

Parkins, J. R., and R. E. Mitchell. "Public Participation as Public Debate: A Deliberative Turn in Natural Resource Management." *Society and Natural Resources* 18, no. 6 (2005): 529–40.

Pateman, C. *Participation and Democratic Theory*. Cambridge: Cambridge University Press, 1970.

Patnaik, S. "Rights against All Odds: How Sacrosanct Is Tribal Forest Rights?" Regional Centre for Development Cooperation, Bhubaneswar, 2008.

Patnaik, U. *The Republic of Hunger and Other Essays*. Gurgaon: Three Essays Collective, 2007.

Patra, H. S., and P. Satpathy. "Legality and Legitimacy of Public Participation in Environmental Decision Making Process: A Review of Scenario from State of Odisha, India." *International Research Journal of Environmental Sciences* 3, no. 3 (March 2014) : 79–84.

Payne, R. A. "Persuasion, Frames and Norm Construction." *European Journal of International Relations* 7, no. 2 (2001): 37–61.

Peters, G. *Institutional Theory in Political Science: The "New Institutionalism."* London: Pinter, 1999.

Phillips, A. *The Politics of Presence.* Oxford: Oxford University Press, 1995.

Pierce, C. S. "Illustrations of the Logic of Science I." *Popular Science Monthly* 12 (November 1877).

Piper, L., and B. Von Lieres. "Expert Advocacy for the Marginalised: How and Why Democratic Mediation Matters to Deepening Democracy in the Global South." IDS working paper, vol. 2011, no. 364. Brighton: Institute of Development Studies at the University of Sussex, 2011.

Pisupati, B. "Economic Value of Biodiversity." *Business Line*, April 9, 2012. www.the-hindubusinessline.com/opinion/economic-value-of-biodiversity/article3297054.ece (accessed June 5, 2012).

Planning Commission. The First Five-Year Plan. Government of India, New Delhi, 1951.

———. The Second Five-Year Plan. Government of India, New Delhi, 1956.

———. The Third Five-Year Plan. Government of India, New Delhi, 1961.

———. The Fourth Five-Year Plan. Government of India, New Delhi, 1969.

———. The Fifth Five-Year Plan. Government of India, New Delhi, 1974.

———. Rolling Plan. Government of India, New Delhi, 1978.

———. The Sixth Five-Year Plan. Government of India, New Delhi, 1980.

———. The Seventh Five-Year Plan. Government of India, New Delhi, 1985.

———. The Eight Five-Year Plan. Government of India, New Delhi, 1992.

———. The Ninth Five-Year Plan. Government of India, New Delhi, 1997.

———. *Report of the Task Force on Panchayati Raj Institutions (PRIs).* New Delhi, December 2001.

———. The Tenth Five-Year Plan. Government of India, New Delhi, 2002.

———. *Status Paper on the Naxal Problem.* 2006. www.satp.org/satporgtp/countries/india/document/papers/06Mar13_Naxal%20Problem%20.htm (accessed June 10, 2012).

———. The Eleventh Five-Year Plan. Government of India, New Delhi, 2007.

———. *Development Challenges in Extremist Affected Areas.* Report of an expert group to Planning Commission. Government of India, New Delhi, 2008.

Poffenberger, M., B. McGean, and A. Khare. "Communities Sustaining India's Forests in the Twenty-First Century." In *Village Voices, Forest Choices: Joint Forest Management in India*, edited by M. Poffenberger and B. McGean, 17–55. New Delhi: Oxford University Press, 1996.

Polletta, F. " 'Free Spaces' in Collective Action." *Theory and Society* 28, no. 1 (1999): 1–38.

———. *Freedom Is an Endless Meeting: Democracy in American Social Movements.* Chicago: University of Chicago Press, 2002.

Prasad, R., and S. Kant. "Institutions, Forest Management, and Sustainable Human Development – Experiences from India." *Environment Development and Sustainability* 5 (2003): 353–67.

Price, V., and J. N. Cappella. "Online Deliberation and Its Influence: The Electronic Dialogue Project in Campaign 2000." *IT and Society* 1 (2002): 303–28. http://citeseerx.ist.psu.edu/viewdoc/download?doi=10.1.1.9.5945&rep=rep1&type=pdf (accessed October 21, 2012).

Putnam, R. *Bowling Alone: The Collapse and Revival of American Community.* New York: Simon & Schuster, 2000.

Qaiyum, N. J. "For Government's Eyes Only." *Down to Earth*, August 15, 1997.

Rai, S. M. "Women and the State in the Third World: Some Issues for Debate." In *Women and the State: International Perspectives*, edited by S. M. Rai and G. Lievesley, 5–22. London: Taylor and Francis, 1996.

Rajan, M., and G. Rajan. *Global Environmental Politics: India and the North-South Politics of Global Environmental Issues*. New Delhi: Oxford University Press, 1997.

Rangan, H. *Of Myths and Movements: Rewriting Chipko into Himalayan History*. New Delhi: Oxford University Press, 2000.

Rao, J. M. "Whither India's Environment?" *Economic and Political Weekly* 30, no. 9 (April 1, 2005): 677–86.

Rao, V., and P. Sanyal. "Dignity through Discourse: Poverty and the Culture of Deliberation in Indian Village Democracies." *Annals of the American Academy of Political and Social Science* 629, no. 1 (2010): 146–72.

Rektor, L. "Advocacy – The Sound of Citizens' Voices." A position paper from the advocacy working group. Ottawa: Government of Canada, Voluntary Sector Initiative Secretariat, 2002.

Rights and Resources. "Operationalising the Act: The Role of Civil Society." www.right sandresources.org/documents/files/doc_962.pdf (accessed October 12, 2011).

Risse, T., C. R. Stephen, and K. Sikkink. *The Power of Human Rights, International Norms and Domestic Change*. New York: Cambridge University Press, 1999.

Risse-Kappen, T., ed. *Bringing Transnational Relations Back In: Non-state Actors, Domestic Structures and International Institutions*. Cambridge: Cambridge University Press, 1995.

Roe, E. *Narrative Policy Analysis: Theory and Practice*. Durham, NC: Duke University Press, 1994.

Rose, N. "Governing Cities, Governing Citizens." In *Democracy, Citizenship and the Global*, edited by E. F. Isin 95–109. London: Routledge, 2000.

Rosencranz, A., and S. Lele. "Supreme Court and India's Forests." *Economic & Political Weekly* 43, no. 5 (February 2, 2008): 11–14.

Rosendal, G. K. *The Convention on Biological Diversity and Developing Countries*. Dordrecht: Kluwer Academic, 2000.

Rosenstone, St. J., and J. M. Hansen. *Mobilization, Participation, and Democracy in America*. New York: Macmillan, 1993.

Rucht, D. "The Strategies and Action Repertoires of New Movements." In *Challenging the Political Order: New Social and Political Movements in Western Democracies*, edited by R. J. Dalton and M. Kuechler, 156–75. New York: Oxford University Press, 1990.

Rudolph, L. I., and Rudolph, S. H. *In Pursuit of Lakshmi: The Political Economy of the Indian State*. Chicago: University of Chicago Press, 1987.

Ryfe, D. M. "The Practice of Deliberative Democracy: A Study of 16 Deliberative Organizations." *Political Communication* 19 (2002): 359–77.

Saberwal, V., and A. Chhatre. "Parvati and the *Tragopan* – Politics, Conservation and Development." *India Together*, April 1, 2002. www.indiatogether.org/environment/arti cles/ghnp/politics.htm (accessed June 12, 2012).

Sahu, S. K. "The Taming of the Wilds." *Infochange*, 2010a. http://infochangeindia.org/ Environment/Community-forests-of-Orissa/The-taming-of-the-wilds.html (accessed March 3, 2011).

———. "What Difference Has the Forest Rights Act Made?" *Infochange*, April 2010. http://infochangeindia.org/environment/community-forests-of-orissa/what-difference-has-the-forest-rights-act-made.html (accessed May 20, 2010).

———. "Converging into a Whole." *Himal* 5 (July 2012): 10.

Salamon, L. M. *The Tools of Government: A Guide to the New Governance.* New York: Oxford University Press, 2002.

Samuel, J. "Public Advocacy in Indian Context." 1989. www.sristi.org/ispe_old/public_advocacy.pdf (accessed April 29, 2004).

Sanders, L. M. "Against Deliberation." *Political Theory* 25 (1997): 347–76.

Sapru, R. K. *Public Policy: Art and Craft of Policy Analysis.* New Delhi: PHI, 2010.

Sarin, M. *From Conflicts to Collaboration: Local Institutions in Joint Forest Management.* Society for Promotion of Wasteland Development. New Delhi: Ford Foundation, 1993.

———. *Joint Forest Management: The Hariyana Experiences.* Ahmedabad: Centre for Environmental Education, 1996.

———. "Comment: Who Is Encroaching on Whose Land?" *Seminar*, November 2002. www.indiaseminar.com/2002/519/519%20comment.html.

———. "Bad in Law." *Down to Earth*, July 15, 2003.

———. "Scheduled Tribes Bill: A Comment." *Economic and Political Weekly* 40, no. 21 (May 5, 2005).

———. "Democratizing India's Forests through Tenure and Governance Reforms." April 10, 2010. www.isidelhi.org.in/saissues/articles/art1apr10.pdf (accessed October 12, 2011).

Sarin, M., and O. Springate-Baginski. "India's Forest Rights Act: The Anatomy of a Necessary but Not Sufficient Institutional Reform." IPPG Discussion Paper 45, University of Manchester, 2010. www.ippg.org.uk/papers/dp45.pdf.Sartori, G. *The Theory of Democracy Revisited: Part One; The Contemporary Debate.* Chatham, NJ: Chatham House Publishers, 1987.

Schmitter, P. C. "Neo-corporatism and the State." In *The Political Economy of Corporatism*, edited by W. Grant 32–63. London: Macmillan, 1985.

Schneider, A. L., and H. Ingram. *Policy Design for Democracy.* Lawrence: University Press of Kansas, 1993.

Scott, D., and C. Barnett. "Something in the Air: Civic Science and Contentious Environmental Politics in Post-apartheid South Africa." *Geoforum* 40, no. 3 (2009): 373–82.

Searles, R., and J. A. Williams. "Negro College Students' Participation in Sit-ins." *Social Forces* 40 (1962): 215–20.

Secretariat of the Convention on Biological Diversity. *Programme of Work on Protected Areas (CBD Programmes of Work).* Montreal: Secretariat of the Convention on Biological Diversity, 2004.

Sekhsaria, P. "Wildlife Is on the Brink." *Hindu Sunday Magazine*, November 1, 2009. www.hindu.com/mag/2009/11/01/stories/2009110150170500.htm (accessed February 2, 2011).

Sethi, H. "Groups in a New Politics of Transformation." *Economic and Political Weekly* 18 (1984): 305–16.

Sethi, N. "Now, an MSP for Forest Produce." *Times of India*, May 1, 2011.

Shah, A., and B. Guru. "Poverty in Remote Rural Areas in India: A Review of Evidence and Issues." Working Paper 21, CPRC and IIPA, 2004.

Shah, C. *Social Movements in India.* New Delhi: Sage Publications, 2004.

Shah, G. *Social Movement in India: A Review of Literature.* Delhi: Sage, 1991.

Shapiro, I. "Enough of Deliberation: Politics Is about Interests and Power." Edited by S. Macedo, 28–38. Oxford: Oxford University Press, 1999.

Sharma, S. D. *Development and Democracy in India.* Boulder: Lynne Rienner Publishers, 1999.

Sherman, R. R., and R. B. Webb. "Qualitative Research in Education: A Focus." In *Qualitative Research in Education: Focus and Methods*, edited by R. R. Sherman and R. B. Webb. London: Falmer Press, 1988.

Sheth, D. L. "Grassroots Initiatives in India." *Economic and Political Weekly* 19, no. 6 (1984): 1–21.

———. "Micro-movements in India: Towards a New Politics of Participatory Democracy." In *Democratizing Democracy: Beyond the Liberal Democractic Canon*, edited by Boaventura de Sousa Santos, 3–37. New York: Verso, 2007.

Shiralkar, K. "80,000 Tribals Demonstrate for Forest Rights Bill." *People's Democracy* 30, no. 33 (August 13, 2006). http://pd.cpim.org/2006/0813/08132006_maha.htm (accessed October 15, 2009).

Shiva, V. *Ecology and Politics of Survival: Conflicts over Natural Resources in India*. New Delhi: Sage Publications, 1991.

Singh, M. "May I Dwell in the Forest?" *Indian Express*, May 7, 2005.

Sinha, P. C. *Green Movements*. New Delhi: Anmol Publications, 1998.

Sivaramakrishnan, K. *Modern Forests: State-Making and Environmental Change in Colonial Eastern India*. Cambridge: Cambridge University Press, 2000.

Sivaramakrishnan, K. "Nationalisms and the writing of environmental histories." *Seminar*, April 19, 2003. www.india-seminar.com/2003/522/522%20k.%20sivaramakrishnan.htm (accessed May 2, 2010).

Skocpol, T., and M. Fiorina, eds. *Civic Engagement in American Democracy*. Washington, DC: Brookings/Russell Sage Foundation, 1999.

Smith, G. *Deliberative Democracy and the Environment*. London: Routledge, 2003.

Smith, G., and C. Wales. "Citizens' Juries and Deliberative Democracy." *Political Studies* 48, no. 1 (2000): 51–65.

Smythies, E. A. *India's Forest Wealth*. 2nd ed. London: Humphrey Milford, 1925.

Snow, D. A., and R. D. Benford. "Ideology, Frame Resonance, and Participant Mobilization." *International Social Movement Research* 1 (1988): 197–217.

Spradley, J. P. *The Ethnographic Interview*. New York: Holt, Rinehart and Winston, 1979.

Squires, J. "Deliberation and Decision Making: Discontinuity in the Two-Track Model." In *Democracy as Public Deliberation: New Perspectives*, edited by M. P. d'Entrèves, 133–56. Manchester: Manchester University Press, 2002.

Srinivas, K. R. "Biodiversity Bill: Nice Words, No Vision." *Economic and Political Weekly* (November 4, 2000): 3916–19.

Stake, R. E. *The Art of Case Study*. London: Sage, 1995.

Stebbing, E. P. *The Forests of India*. Vol. 2. London: Bodley Head, 1923.

Stiefel, M., and M. Wolfe. "Participation in the 1990s." *A Voice for the Excluded: Popular Participation in Development; Utopia or Necessity*, chap. 10. Geneva/London: UNRISD/Zed Books, 1994.

Stivers, C. "Toward Administrative Public Space: Hannah Arendt Meets the Municipal Housekeepers." *Administration & Society* 34, no. 1 (2002): 98–102.

Stone, D. *Policy Paradox*. New York: W. W. Norton, 1988.

———. *Policy Paradox and Political Reason*. Glenview, IL: Scott Foresman, 1998.

———. *Policy Paradox: The Art of Political Decision Making*. Rev. ed. New York: W. W. Norton, 2002.

Subotic, J. *Hijacked Justice: Dealing with the Past in the Balkans*. Ithaca: Cornell University Press, 2009.

Sundar, N. "Violent Social Conflicts in India's Forests – Society, State and the Market." Conference on Indian Forestry: Key Trends and Challenges, New Delhi, March 5–6, 2009.

Sundar, N., R. Jeffery, and N. Thin. *Branching Out: Joint Forest Management in India.* New Delhi: Oxford University Press, 2001.

Swiderska, K. "Stakeholder Participation in Policy on Access to Genetic Resources, Traditional Knowledge and Benefit-Sharing, Case Studies and Recommendations." *Biodiversity and Livelihoods Issues*, no. 4 (March 2001): 1–35.

Tandon, R. "Linking Citizenship, Participation and Accountability: A Perspective from PRIA." IDS Bulletin 33, April 2, 2002.

Tandon R., and R. Mohanty. *Civil Society and Governance.* New Delhi: Pria, 2002.

Taneja, B., and A. Kothari. "Indian Case Study." In *Biodiversity Planning in Asia* edited by J. Carew-Reid 369–406. Sri Lanka: IUCN, 2002.

Tarrow, S. G. *Struggle, Politics, and Reform: Collective Action, Social Movements, and Cycles of Protest.* Ithaca, NY: Center for International Studies, Cornell University, 1989.

———. *Power in Movement.* New York: Cambridge University Press, 1994.

———. *Power in Movement: Social Movements and Contentious Politics.* Cambridge: Cambridge University Press, 1998.

Thakuria, N., S. Joshi, and S. Barik. "The Battle over Forests." *Down to Earth*, January 15, 2003. www.downtoearth.org.in/node/12366 (accessed June 17, 2009).

Timsina, N. "Empowerment or Marginalization: A Debate in Community Forestry in Nepal." *Journal of Forest and Livelihood* 2 (2002): 27–33.

Toke, D., and D. Marsh. "Policy Networks and the GM Crops Issue: Assessing the Utility of a Dialectical Model of Policy Networks." *Public Administration* 81 (2003): 229–51.

Torgerson, D. *The Promise of Green Politics.* Durham, NC: Duke University Press, 1999.

Touraine, A. "An Introduction to the Study of Social Movements." *Social Research* 52, no. 4 (1985): 749–87.

Upadhyay, S. "The Challenges of Access to Justice for the Marginalised in India: The Law and the Process; The Case of Forest Rights Presentation." CNRS, Paris 2003. Presentation for JUST-INDIA, 2011.

Van Manen, M. "Linking Ways of Thinking with Ways of Being Practical." *Curriculum Inquiry* 6, no. 3 (1977): 205–29.

Venkatesan, J. "Order on Encroachments on Forest Lands Stayed." *Hindu*, February 23, 2004.

Verba, S., K. L. Schlozman, and H. E. Brady. *Voice and Equality: Civic Voluntarism in American Politics.* Cambridge: Harvard University Press, 1995.

Wainwright, H. *Arguments for a New Left: Answering the Free-Market Right.* Oxford: Blackwell, 1994.

Walzer, M. "Deliberation, and What Else?" In Deliberative Politics. Essays on Democracy and Disagreement. Edited by S. Macedo, 58–69. Oxford: Oxford University Press, 1999.

Warren, M. E. *Democracy and Association.* Princeton: Princeton University Press, 2001.

Weber, E. P. *Bringing Society Back In: Grassroots Ecosystem Management, Accountability, and Sustainable Communities.* Cambridge, MA: MIT Press, 2003.

Wight-Felske, A. "History of Advocacy Tool Kit." In *Making Equality: History of Advocacy and Persons with Disabilities in Canada*, edited by D. Stienstra and A. Wight-Felske, 321–38. Concord, ON: Captus Press, 2003.

Williams, B. A., and A. R. Matheny. *Democracy, Dialogue, and Environmental Disputes: The Contested Languages Of Social Regulation.* New Haven: Yale University Press, 1995.

Williams, G., and Mawdsley, E. "Postcolonial Environmental Justice: Government and Governance in India." *Geoforum* 37, no. 5 (2006): 660–70.

Wittmer, H., and R. Birner. "Between Conservationism, Eco-populism and Developmentalism – Discourses in Biodiversity Policy in Thailand and Indonesia." CAPRi Working

Paper 37. IFPRI, Washington, DC, 2005. www.ifpri.org/sites/default/files/publications/CAPRiWP37.pdf.

World Bank. *Unlocking Opportunities for Forest Dependant People*. New Delhi: World Bank, 2006.

World Climate Conference 3. Geneva, Switzerland, 2009.

Wuthnow, R. *Sharing the Journey: Support Groups and America's New Quest for Community*. New York: Free Press, 1994.

Yadav, Y. "The Elusive Mandate of 2004." *Economic and Political Weekly* (December 18, 2004): 5383–5398.

Yanow, D. "Silences in Public Policy Discourse: Organizational and Policy Myths." *Journal of Public Administration Research and Theory* 2, no. 4 (1992): 399–423.

———. "Built Space as Story." *Policy Studies Journal* 23 (1995): 407–22.

———. *How Does a Policy Mean? Interpretive Policy and Organizational Actions*. Washington, DC: Georgetown University Press, 1996.

———. *Conducting Interpretive Policy Analysis*. Thousand Oaks, CA: Sage Publications, 2000.

Yin, R. K. *Case Study Research: Design and Methods*. 3rd ed. Thousand Oaks, CA: Sage, 2003.

———. *Case Study Research: Design and Methods*. 4th ed. Thousand Oaks, CA: Sage, 2009.

Young, I. M. *Justice and the Political Difference*. Princeton: Princeton University Press, 1990.

———. "Activist Challenges to Deliberative Democracy." *Political Theory* 29 (2001): 670–90.

———. "Activist Challenges to Deliberative Democracy." In Philosophy, Politics and Society 7: Debating Deliberative Democracy. Edited by J. Fishkin and P. Laslett, 102–120. Oxford: Blackwell, 2003.

Young, K. " 'Values' in the Policy Process." *Policy and Politics* 5 (1977): 1–22.

Zainol, Z. A., L. Amin, F. Akpoviri, and R. Ramli. "Biopiracy and States' Sovereignty over Their Biological Resources." *African Journal of Biotechnology* 10, no. 58 (2011): 12395–408. www.academicjournals.org/ajb/pdf.

Zald, M. N. "Ideologically Structured Action: An Enlarged Agenda for Social Movement Research." *Mobilization* 5, no. 1 (2000): 1–16.

Zolo, D. *Democracy and Complexity: A Realist Approach*. University Park: Pennsylvania State Press, 1992.

Index